# Update on Orthoflavivirus - Understanding Orthoflavivirus, Potential Reemerging Pathogens

*Edited by José Antonio Morales-Gonzalez,*
*Jazmín García-Machorro*
*and Gabriela Mellado-Sánchez*

Published in London, United Kingdom

Update on Orthoflavivirus - Understanding Orthoflavivirus, Potential Reemerging Pathogens
http://dx.doi.org/10.5772/intechopen.1008608
Edited by José Antonio Morales-Gonzalez, Jazmín García-Machorro and Gabriela Mellado-Sánchez

Contributors
Aitana Juan-Giner, Alejandra García-Aguilar, Ana Carolina Campi-Azevedo, Andreza Parreiras Gonçalves, Andréa Teixeira-Carvalho, Daniel Duran-Ayala, Derick Kimathi, Desiree Izar-Marín, Esther Candelaria Martínez-Molina, Gabriela Mellado-Sánchez, Gamou Fall, George Warimwe, Giovani Visoso-Carvajal, Izabela Maurício de Rezende, Jaquelline Germano De Oliveira, Jazmín García-Machorro, Jordana Rodrigues Barbosa Fradico, Josselin Carolina Corzo-Gómez, José Antonio Morales-González, Juan Carlos Muñoz-Herrera, Mara Gutiérrez-Sánchez, Martha Pedraza-Escalona, María Fernanda García-Aguilar, Mauricio Vázquez-Pichardo, Mayra Montecillo-Aguado, Moussa Dia, Ndeye Sakha Bob, Ndongo Dia, Octavio Rodríguez-Cortés, Olindo Assis Martin-Filho, Oumar Faye, Raúl Flores-Mejía, Rebecca F. Grais, Samantha Climaco-Arvizu, Sandra Paola Martínez-Frías, Sonia Mayra Pérez-Tapia, Sury Antonio López-Cancino, Victor Javier Cruz-Holguin, Ángel Adrián Jiménez-Vega



First published in London, United Kingdom, 2025 by IntechOpen
IntechOpen is the global imprint of INTECHOPEN LIMITED, registered in England and Wales, registration number: 11086078, 167-169 Great Portland Street, London, W1W 5PF, United Kingdom

For EU product safety concerns: IN TECH d.o.o., Prolaz Marije Krucifikse Kozulić 3, 51000 Rijeka, Croatia, info@intechopen.com or visit our website at intechopen.com.

British Library Cataloguing-in-Publication Data
A catalogue record for this book is available from the British Library

Update on Orthoflavivirus - Understanding Orthoflavivirus, Potential Reemerging Pathogens
Edited by José Antonio Morales-Gonzalez, Jazmín García-Machorro and Gabriela Mellado-Sánchez
p. cm.
Print ISBN 978-1-83635-460-4
Online ISBN 978-1-83635-459-8
eBook (PDF) ISBN 978-1-83635-461-1

If disposing of this product, please recycle the paper responsibly.

# Meet the editors

José Antonio Morales-González graduated as a surgeon physician from FES-Iztacala, National Autonomous University of Mexico (UNAM). He engaged in doctoral studies in biological sciences at the UNAM. Dr. Morales-González has been awarded diverse recognitions: the Gustavo Baz Prada Medal by the UNAM and the Alfonso Caso Medal for academic merit by the UNAM, and was distinguished by the National System of Researchers (SNI) as National Researcher Level 3 (2026–2030). He is the author of 120 internationally published articles, which have received more than 6,000 citations. He is also the editor and coordinator of 33 specialized books and the thesis director of 16 undergraduate students and 120 graduate students. Guest editor of six special issues in international journals. Dr. Morales-González is a full-time tenured professor-researcher at the Escuela Superior de Medicina, Instituto Politécnico Nacional in Mexico.

Jazmín García-Machorro graduated as a surgeon physician from the Escuela Superior de Medicina (ESM) of the National Polytechnic Institute (IPN). She pursued a master's degree and a doctorate in Molecular Biomedicine at CINVESTAV. The IPN awarded Dr. García-Machorro the Applied Research Award in December 2023. Additionally, in 2019, she obtained the CANIFARMA Award, third place in the technological research category. She was distinguished by the National System of Researchers (SNI) as National Researcher Level 2 (2026-2030), Member of the Mexican Society of Virology, Member of the Mexican Biosafety Association, and Member of the Salud IPN network. She is the author of 35 internationally published articles. She is a full-time tenured professor-researcher at the ESM, IPN.

Gabriela Mellado-Sánchez graduated as a Chemist, Bacteriologist, and Parasitologist, with Master's and Doctorate studies in Immunology, obtained at the National School of Biological Sciences (ENCB) of the National Polytechnic Institute (IPN). She got a postdoctoral position at the University of Maryland, Baltimore, USA. Dr. Mellado Sánchez was awarded the Roche Syntex Dr. Rosenkranz Award 2008 in Medical Research, Basic area, and was distinguished by the National System of Researchers (SNI) as National Researcher Level 2 (2024–2028). She has published 30 scientific articles (more than 370 citations, h-index = 12) as well as 10 chapters in scientific books. She has worked in academic and industrial settings related to biologics. Currently, she serves as the Sanitary Compliance Officer for the Biotherapeutics Research and Development Unit (UDIBI) at ENCB-IPN.

# Contents

# Preface

Conservation medicine studies the interrelationship among the environment, animal health, and human health. Within these interconnections, it explores bidirectionally the processes of health and disease, as well as the impact of ecosystems that may drive certain pathogens to cross the species barrier, expanding transmission to new hosts. In this sense, conservation medicine addresses zoonoses, which are defined as infectious diseases that are naturally transmitted from animals to humans, either directly, indirectly, or through an intermediate host.

Among these intermediates—also known as vector-borne transmission—are insects and other arthropods (vectors) such as mosquitoes or ticks, which transmit pathogens from animals to humans or vice versa, while incorporating this process into their life cycle. Viruses are among the most extensively studied pathogens that have caused serious public health problems, as they are constantly evolving due to the nature of their genomes and the impact of ecosystem changes. This underscores the need for continuous knowledge updates, which gave rise to the present book.

In April 2023, the International Committee on Taxonomy of Viruses (ICTV) approved a new taxonomy for the family *Flaviviridae*. The genus name *Flavivirus* was changed to *Orthoflavivirus* to avoid ambiguities with the root "flavi", which had been present both in the family and genus. In addition, all species names within this family were modified to a binomial format, similar to the nomenclature used for other biological species. For example, what was previously known as "Dengue virus" is now designated as *Orthoflavivirus denguei*. Thus, the *Flaviviridae* family currently includes four genera (*Hepacivirus, Orthoflavivirus, Pegivirus*, and *Pestivirus*) and 89 species. This book focuses on *Orthoflavivirus*, providing updated information on nomenclature, recent discoveries related to transmission, immune responses, novel treatments, diagnosis, and prognosis in humans. It also addresses viruses of veterinary importance transmitted by vectors or between rodents and other animals, such as bats. Special emphasis is placed on preventive and control measures, ranging from personal protection to reducing vector habitats.

*Update on Orthoflavivirus – Understanding Orthoflavivirus, Potential Reemerging Pathogens* is composed of seven highly specialized chapters covering topics such as the evaluation and development of vaccines through clinical trials using PRNT for "Yellow Fever Virus" (*Orthoflavivirus flavi*). One chapter addresses the challenges and advances of vaccines and immune responses to *Orthoflavivirus*. Another chapter presents an in-depth discussion on *Orthoflavivirus denguei*, focusing on humoral immune responses, vertical transmission, and coinfection by *Aedes aegypti*—the latter also including "Zika virus" (*Orthoflavivirus zikaense*). Other specialized topics include the impact of obesity on clinical severity caused by *Orthoflavivirus* infection and an updated review of hepatic damage associated with severe dengue, in which alcohol consumption is described as a risk factor that exacerbates liver injury induced by *Orthoflavivirus denguei*.

This book also serves as a guide to facilitate the transition from the former nomenclature to the newly established taxonomy of this viral family. This book was authored by researchers from diverse international academic institutions (Mexico, France, Dakar, and Kenya), underscoring its scientific quality. We hope that this editorial effort, led by the Laboratory of Conservation Medicine of the Escuela Superior de Medicina, Instituto Politécnico Nacional, meets and exceeds your expectations.


**Dra. Jazmín García-Machorro and Dr. José Antonio Morales-González**<br>
Laboratorio de Medicina de la Conservación,<br>
Escuela Superior de Medicina,<br>
Instituto Politécnico Nacional,<br>
Mexico City, Mexico

**Dra. Gabriela Mellado-Sánchez**<br>
Unidad de Desarrollo e Investigación en Bioterapéuticos (UDIBI),<br>
Escuela Nacional de Ciencias Biológicas,<br>
Instituto Politécnico Nacional,<br>
Mexico City, Mexico

Chapter 1

# Introductory Chapter: From the First Described Flavivirus to the Latest Classification

*Gabriela Mellado-Sánchez, José Antonio Morales-González and Jazmín García-Machorro*

## 1. Introduction

### 1.1 Flaviviruses: A persistent threat

Flaviviruses (from the Latin "flavus," meaning "yellow") are small, spherical viruses measuring approximately 50 nanometers in diameter. Their genetic material consists of a single-stranded, positive-sense RNA genome, ranging from 9 to 13 kilobases in length. The virus structure includes a protein capsid (C) with icosahedral symmetry and a lipid envelope containing two key proteins: the envelope (E) and membrane (M) proteins [1].

These viruses primarily infect mammals—including humans, pigs, horses, cattle, sheep, goats, rodents, and primates—but they have also been found in birds, reptiles, and even bats, which serve as natural reservoirs. Transmission occurs in both wild (sylvatic) and urban environments through arthropod vectors, such as mosquitoes and ticks (see **Figure 1**).

In humans, flavivirus infections can cause a wide range of diseases, from fever and skin rashes to hemorrhagic syndromes, neurological complications, severe liver damage, and even death.

### 1.2 A bit of history

The first virus ever identified in humans was the yellow fever virus (YFV), discovered in 1901. Its name refers to the characteristic yellowing of the skin and mucous membranes (jaundice), caused by liver damage and elevated bilirubin levels in the blood. This is also the origin of the name of the *Flaviviridae* family [1].

Symptoms of dengue virus (DENV) infection were described in China during the Jin Dynasty (265–420 AD). In the Americas, the first reported outbreaks occurred in the eighteenth and nineteenth centuries. The viral nature of dengue was confirmed in the 1940s, and the first DENV serotypes were isolated in 1943 [2].

The Zika virus (ZIKV) was first discovered in 1947 in the Zika Forest in Uganda. Since then, more viruses belonging to this family have been identified. Interestingly, flaviviruses were initially classified under the *Togaviridae* family due to structural

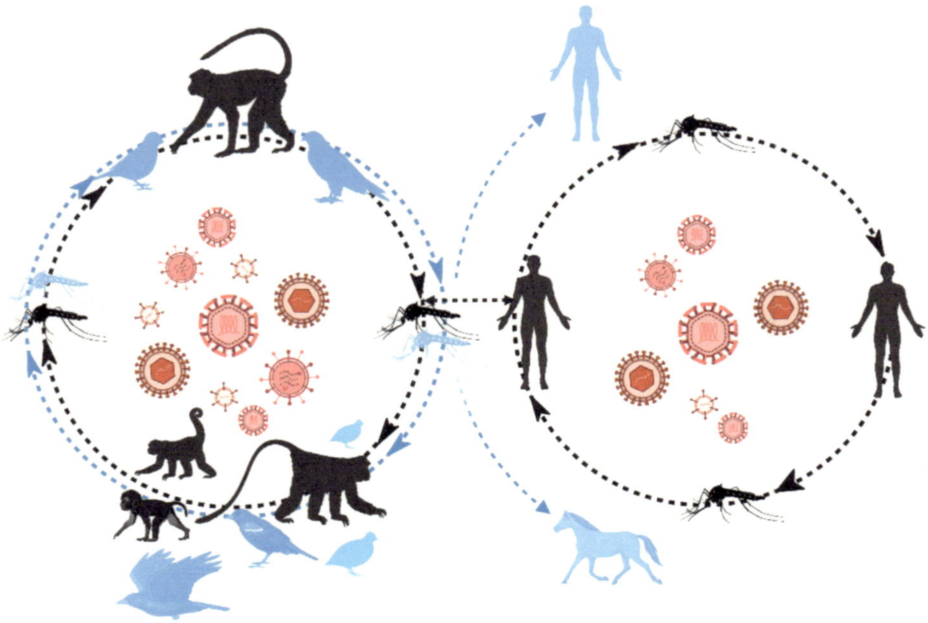

**Figure 1.**
*Sylvatic and urban transmission cycles of flaviviruses. In black, viruses such as yellow fever (YFV), dengue (DENV), and Zika (ZIKV); in blue, others such as West Nile virus (WNV) and Japanese Encephalitis virus (JEV).*

similarities with alphaviruses. It was not until 1984 that the *Flaviviridae* family was established, based on key differences in genome structure and viral replication mechanisms [3].

### 1.3 Expanding vectors

One of the major public health challenges today is the geographical expansion of flavivirus vectors, particularly the mosquitoes *Aedes aegypti* and *Aedes albopictus*. Deforestation, urban expansion, and climate change have enabled these vectors to inhabit new areas, increasing the risk of outbreaks in previously unaffected regions.

A clear example is ZIKV, which originally circulated only in Africa and Asia. In 2015, it spread to Central and South America, becoming a reemerging disease in the region [4]. Since then, it has circulated alongside other flaviviruses such as DENV, which was already considered a reemerging disease in tropical areas.

## 2. What this book covers

This book explores various scientific and clinical aspects of flaviviruses, beginning with YFV, one of the oldest and most well-known viruses. Its vaccine remains among the most effective and long-lasting, making it important to understand how its efficacy is evaluated in the laboratory.

We then delve into the molecular and immunological basis of the antibody response induced by dengue virus (DENV), a particularly complex pathogen with four antigenically distinct serotypes. The book also addresses cross-reactivity with other flaviviruses, such as ZIKV, and the phenomenon known as antibody-dependent enhancement (ADE), which can worsen disease outcomes.

Additionally, a chapter is dedicated to recent advances in vaccine platforms being developed for this virus family—a highly relevant topic for global health.

Other chapters explore the link between DENV and obesity, as obesity may increase the risk of developing severe complications. We also examine liver damage caused by DENV, discussing potential pathological mechanisms and proposing clinical and biochemical liver monitoring as a tool for prognosis.

The book concludes with a chapter on the use of botanical extracts as an alternative strategy to control mosquito vectors of DENV and ZIKV. This connects with the principles of conservation medicine, which studies the interrelationships between environmental, human, and animal health—the central theme of this book.

## 3. A new name for familiar viruses

On the other hand, this work seeks to disseminate the most current nomenclature of flaviviruses, given their relevance in public health. In 2023, the International Committee on Taxonomy of Viruses (ICTV) officially replaced the genus name *Flavivirus* with *Orthoflavivirus*, while retaining the *Flaviviridae* family name [5]. This change aims to avoid confusion between family and genus names and to implement a binomial system for naming viral species.

Some examples of the new nomenclature include:

- YFV → *Orthoflavivirus flavi*
- DENV → *Orthoflavivirus denguei*
- ZIKV → *Orthoflavivirus zikaense*
- West Nile virus → *Orthoflavivirus nilense*
- Japanese encephalitis virus → *Orthoflavivirus japonicum*

## 4. Final thoughts

Even with improved classification systems and effective vaccines for some flaviviruses, if sylvatic transmission cycles continue to exist, these viruses will remain difficult to eliminate. Infected vectors will continue to facilitate transmission to humans and other vertebrate hosts, making orthoflavivirus a continuing concern for both human and animal health.

## Acknowledgements

The authors gratefully acknowledge CONAHCYT, Instituto Politécnico Nacional and SIP/IPN grants: 20250361, 20254045.

## Conflict of interest

The authors declare no conflict of interest.

## Nomenclature

| | |
|---|---|
| ADE | antibody-dependent enhancement |
| C | capsid |
| DENV | dengue virus, *Orthoflavivirus denguei* |
| E | envelope |
| JEV | Japanese encephalitis virus, *Orthoflavivirus japonicum* |
| M | membrane |
| WNV | West Nile virus, *Orthoflavivirus nilense* |
| YFV | yellow fever virus, *Orthoflavivirus flavi* |
| ZIKV | Zika virus, *Orthoflavivirus zikaense* |

## Author details

Gabriela Mellado-Sánchez[1,2*], José Antonio Morales-González[3] and Jazmín García-Machorro[3*]

1 Unidad de Desarrollo e Investigación en Bioterapéuticos (UDIBI), Escuela Nacional de Ciencias Biológicas, Instituto Politécnico Nacional, Mexico City, Mexico

2 Laboratorio Nacional para Servicios Especializados de Investigación, Desarrollo e Innovación (I+D+i) para Farmoquímicos y Biotecnológicos, LANSEIDI-FarBiotec-CONAHCYT, Mexico City, Mexico

3 Laboratorio de Medicina de la Conservación, Escuela Superior de Medicina, Instituto Politécnico Nacional, Mexico City, Mexico

*Address all correspondence to: melladosg@gmail.com and jazzgama81@gmail.com

## References

[1] Reed W, Carroll J, Agramonte A. The etiology of yellow fever. An additional note. JAMA. 2025;**XXXVI**(7):431-440. Available from: https://jamanetwork.com/journals/jama/article-abstract/469979 [Accessed: September 5, 2025]

[2] Fundación io. Dengue [Internet]. 2025. Available from: https://fundacionio.com/salud-io/enfermedades/virus/dengue/#:~:text=Historia,un%20t%C3%ADpico%20dolor%20de%20huesos [Accessed: September 7, 2025]

[3] Schmaljohn AL, McClain D. Alphaviruses (togaviridae) and flaviviruses (flaviviridae). In: Medical Microbiology. 4th ed. University of Texas Medical Branch at Galveston; 1996

[4] White MK, Wollebo HS, David Beckham J, Tyler KL, Khalili K. Zika virus: An emergent neuropathological agent. Annals of Neurology. 2016;**80**(4):479-489. DOI: 10.1002/ana.24748

[5] Postler TS, Beer M, Blitvich BJ, Bukh J, de Lamballerie X, Drexler JF, et al. Renaming of the genus *Flavivirus* to *Orthoflavivirus* and extension of binomial species names within the family *Flaviviridae*. Archives of Virology. 2023;**168**(9):224. DOI: 10.1007/s00705-023-05835-1

**Chapter 2**

# Yellow Fever: New Insights on an Old Disease

*Jordana Rodrigues Barbosa Fradico,*
*Izabela Maurício de Rezende, Andreza Parreiras Gonçalves,*
*Jaquelline Germano De Oliveira, Ana Carolina Campi-Azevedo,*
*Olindo Assis Martin-Filho and Andréa Teixeira-Carvalho*

## Abstract

Yellow fever (YF) is an acute arboviral hemorrhagic disease caused by the yellow fever virus (YFV), species *Orthoflavivirus flavi*, of the Flaviviridae family. YF remains a significant public health threat in endemic regions of Africa and South America, where recurrent outbreaks have occurred, despite the availability of a safe and effective vaccine. In recent years, YF has reemerged in previously unaffected areas, including densely populated regions of Brazil, highlighting ecological shifts, vaccine coverage gaps, and increased vulnerability of populations, raising concerns about its potential global impact. This chapter provides an updated and comprehensive overview of YF, addressing virological characteristics, transmission cycles, epidemiological trends, and pathogenesis. Special emphasis is placed on recent scientific findings, including sylvatic transmission of YFV in Brazil, which has played a central role in recent outbreaks and presents a continued risk of reurbanization. The clinical spectra of YF are reviewed, ranging from mild febrile illness to severe, multisystem disease associated with high case-fatality rates. Diagnostic strategies are discussed, including molecular and serological tools, as well as biomarkers associated with disease severity. Particular attention is given to the host immune response, including cytokine dysregulation, which contributes to the pathogenesis and outcomes of severe cases. Recent findings on immune-mediated mechanisms of late-relapsing hepatitis are also discussed. Finally, prevention strategies are addressed, with a focus on the safety and efficacy of the vaccine 17D/17DD and the occurrence of rare adverse events following vaccination. The chapter underscores the importance of sustained surveillance strategies, immunization, and research to mitigate the burden of this reemerging disease.

**Keywords:** yellow fever, sylvatic transmission, epidemiology, immunopathogenesis, vaccination, biomarkers, diagnosis

## 1. Introduction

Yellow fever (YF) is a reemerging viral hemorrhagic disease that continues to pose a significant public health challenge in endemic regions of Africa and South America.

Caused by the yellow fever virus (YFV), the disease is transmitted primarily through mosquito bites and may range in presentation from a mild, self-limiting febrile illness to a severe and potentially fatal hemorrhagic condition involving multiple organ failure [1, 2].

Despite the availability of an effective vaccine for more than 80 years, YF outbreaks have increased in recent decades, particularly in unvaccinated populations, owing to uncontrolled urbanization at the forest–urban interface, deforestation, ecological changes, and gaps in immunization coverage [1, 3]. The 2016–2018 outbreaks in Brazil and Angola underscored the disease's epidemic potential and the global risk it represents due to increased human mobility and the presence of competent mosquito vectors in non-endemic areas [3, 4]. The recent outbreaks in South America in 2024–2025 show a change in the geographic distribution of the disease and raise concerns again about the potential of the YFV to establish an urban cycle in this region [5].

This chapter provides an updated overview of YF, including virological and epidemiological features, transmission dynamics, clinical and pathological aspects, host immune responses, diagnostic tools, management strategies, and vaccine-related issues, emphasizing recent scientific findings and public health concerns.

## 2. Yellow fever virus

The YFV, species Ortoflavivirus flavi, of the Flaviviridae family, is a single-stranded RNA virus presented as an enveloped virus with an isometric capsid and a genome of approximately 11 kilobases. Its genome encodes a single polyprotein that is subsequently cleaved into three structural proteins (capsid [C], pre-membrane/membrane [M], and envelope [E]) and seven non-structural (NS) proteins (NS1, NS2A, NS2B, NS3, NS4A, NS4B, and NS5) involved in viral replication, pathogenesis, and immune evasion [2, 6–11].

Phylogenetic analyses have identified seven YFV genotypes—West Africa I, West Africa II, East Africa, Central Africa, Angola, and South America I and II [12]. In South America, two main lineages, from the South American I genotype, circulate, but recent outbreaks in Brazil (2016–2018) revealed a new lineage expanding into southeastern regions previously free of the virus. This spread underscores YFV's potential for ecological shifts and geographical expansion [4].

## 3. Transmission cycle

YFV transmission occurs through complex ecological interactions involving non-human primates (NHPs), mosquitoes, and humans. Three main transmission cycles are recognized: sylvatic (forest), intermediate (savanna/peri-urban forests), and urban (**Figures 1** and **2**). These cycles differ in terms of geographical context, vector species, and host involvement but share the mosquito-borne transmission mechanism [2, 9], as described below.

### 3.1 Sylvatic cycle

In the sylvatic (or enzootic) cycle (**Figures 1** and **2**), YFV circulates between NHPs as primary reservoir hosts and arboreal mosquitoes such as *Haemagogus* spp. and

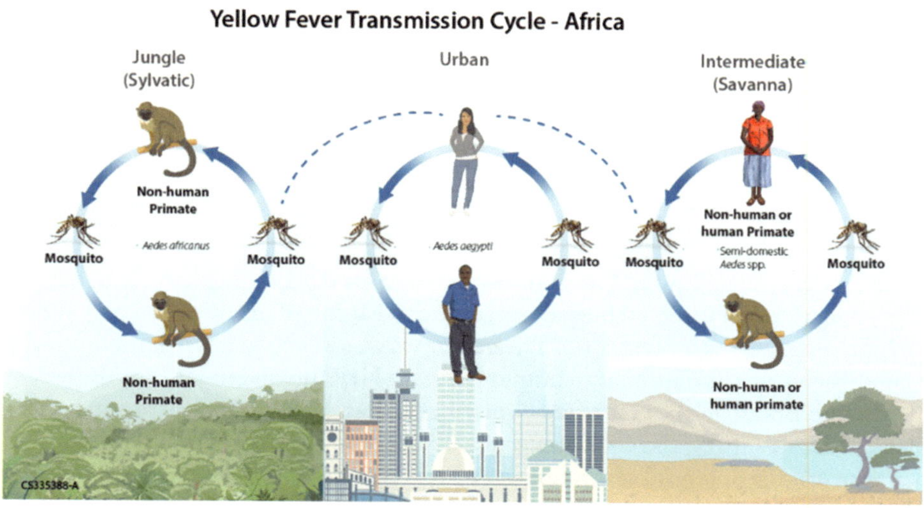


**Figure 1.**
*Yellow fever virus transmission cycles in Africa. Source: CDC. https://www.cdc.gov/yellow-fever/php/transmission/index.htm*

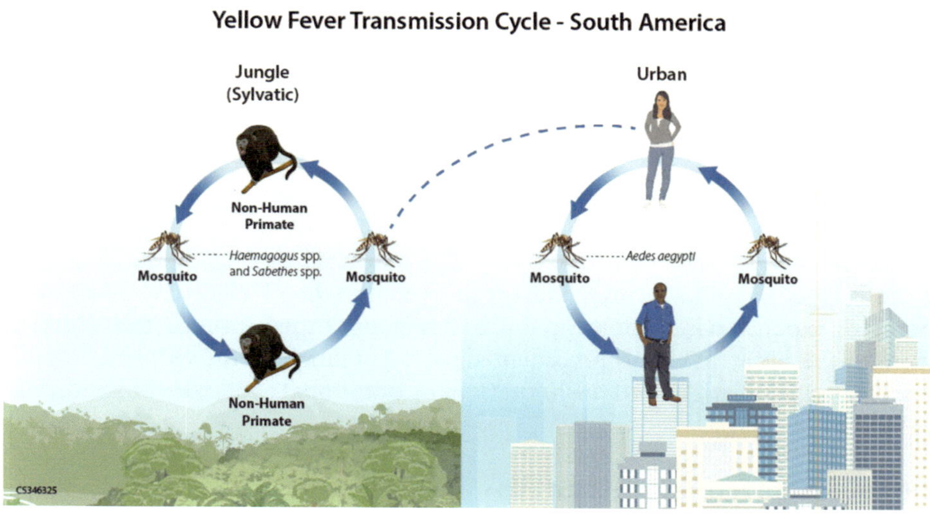


**Figure 2.**
*Yellow fever virus transmission cycles in South America. Source: CDC. https://www.cdc.gov/yellow-fever/php/transmission/index.html*

*Sabethes* spp. as vectors, primarily in South America. Humans can become incidental hosts when venturing into forested areas and are bitten by infected mosquitoes [4, 9].

Recent Brazilian outbreaks (2016–2018) revealed an alarming reemergence of the sylvatic cycle in fragmented forests close to urban areas. Studies have detected viral RNA in NHP carcasses and vector pools in peri-urban forests, emphasizing the increasing risk of spillover to human populations and potential reurbanization of YF [4].

### 3.2 Intermediate cycle (Africa)

This cycle (**Figures 1**), observed in African savanna regions, involves both NHPs and humans as amplifying hosts. Semi-domestic mosquitoes (*e.g., Aedes* spp.) breed near human settlements, facilitating localized outbreaks when conditions are favorable [13].

### 3.3 Urban cycle

The urban cycle (**Figures 1** and **2**), responsible for devastating historical outbreaks, occurs when infected humans introduce YFV into densely populated areas with high *Aedes aegypti* infestation. The vector thrives in urban environments and can transmit the virus from human to human without NHP involvement [9, 13].

Although Brazil has not reported confirmed urban YFV transmission since the mid-twentieth century, the coexistence of YFV circulation and abundant *A. aegypti* raises concern for potential reurbanization, especially in under-vaccinated areas [4].

Several environmental and social factors critically influence the YF transmission dynamics. Climate change and deforestation play a pivotal role by altering ecosystems, promoting the dispersion of sylvatic mosquito vectors, and forcing NHPs into closer contact with human populations. Simultaneously, urban expansion has intensified human-vector interactions in peri-urban areas, where fragmented forest environments overlap with densely populated communities. These conditions create ideal scenarios for zoonotic spillover and potential reurbanization of the virus. Compounding these risks, vaccine coverage gaps—particularly in regions previously considered low-risk—have allowed outbreaks to reach susceptible and immunologically naive populations, often with devastating consequences [3, 4, 6, 14].

## 4. Sylvatic yellow fever in Brazil

Brazil has experienced recurrent sylvatic YF outbreaks throughout its history, but the 2016–2018 epidemic marked a turning point in the YFV's behavior, with its geographic expansion into densely populated regions of the Southeast and South regions of the country. This unprecedented spread highlighted the vulnerability of unvaccinated populations and the limitations of traditional containment strategies [4, 15].

### 4.1 Reemergence and geographical expansion

Historically, YFV circulation was restricted to the Amazon Basin in the Americas region. However, beginning in late 2016, the virus spread rapidly in Brazil through the states of Minas Gerais, Espírito Santo, Rio de Janeiro, and São Paulo—regions with low historical exposure and low vaccine coverage at the time of these outbreaks. Phylogenetic analyses confirmed that the outbreaks were caused by the South American genotype I lineage, which underwent silent enzootic expansion before causing explosive epizootics and human cases [4, 15, 16].

NHP mortality, particularly among *Alouatta* spp. (howler monkeys), served as an early indicator of YFV circulation. Surveillance systems that incorporated primate epizootics were critical for identifying viral hotspots and directing emergency vaccination campaigns [4, 16].

### 4.2 Vectors and environmental drivers

The primary vectors in Brazil's YF sylvatic cycle are mosquitoes of the genera *Haemagogus* and *Sabethes*. These mosquitoes breed in forest canopies but can also be found in forest fragments within urban and peri-urban zones, facilitating zoonotic spillover events. Environmental degradation, including deforestation and forest fragmentation, has played a significant role in increasing contact between humans, NHPs, and sylvatic vectors [4].

### 4.3 Demographic and temporal patterns

Epidemiological data from recent YF outbreaks in Brazil have revealed distinct demographic and seasonal trends. Adults, particularly males aged 20–59 years, are disproportionately affected, likely due to occupational exposure in forested and peri-forested areas, where sylvatic transmission predominates. The disease also follows a marked seasonal pattern, with incidence peaking during the tropical rainy season (December to May), which coincides with increased mosquito breeding and activity [4]. Moreover, case-fatality rates among confirmed cases have ranged from 30 to 50% in severe presentations, underscoring the virulence of the virus and the challenges posed by limited access to intensive care units in many affected rural and remote areas [6].

### 4.4 Public health implications

The reemergence of sylvatic YF in southeastern Brazil revealed gaps in routine immunization and surveillance in areas previously not considered at risk. The outbreaks prompted mass vaccination campaigns, deployment of mobile labs, and intensified entomological monitoring. Importantly, these events reignited the global discussion on the risk of urban YF reestablishment due to the presence of *Aedes aegypti* in affected regions [4, 15].

The outbreaks also demonstrated the importance of integrating eco-epidemiological data—such as NHP surveillance, entomological mapping, and land-use changes—into real-time public health decision-making. Efforts like the Brazilian National Surveillance Plan for Yellow Fever Epizootics in Non-Human Primates (Plano de Vigilância de Epizootias em Primatas Não Humanos) were strengthened to provide early warnings for potential human cases and inform vaccine deployment strategies [4, 16].

Despite control efforts, the continued circulation of YFV in the Atlantic Forest biome and the expansion of competent mosquito vectors to urban fringes indicate an ongoing risk of viral reemergence in previously unaffected areas, especially in the context of low vaccine uptake or delayed immunization response.

## 5. Epidemiological features

YF remains a significant public health concern in tropical regions of Africa and South America, with periodic outbreaks driven by complex ecological, social, and immunological factors. Despite the challenges in surveillance systems leading to underreporting, it is estimated that up to 200,000 cases and 30,000 deaths occur annually worldwide [6]. The disease's epidemiology is characterized by its distinct transmission cycles and the dynamic interplay between human populations, NHPs, and mosquito vectors.

Historically, the urban transmission cycle was responsible for large-scale epidemics in cities across the Americas and Africa during the nineteenth and early twentieth centuries. The successful implementation of vector control programs and mass vaccination campaigns interrupted urban transmission in the Americas by the mid-twentieth century. However, the sylvatic cycle persists in forested areas, maintaining an ecological reservoir in NHPs and mosquitoes such as *Haemagogus sp.* and *Sabethes spp.* in South America and *Aedes africanus* in Africa [17].

## 5.1 Global burden and distribution

Endemic transmission persists in different African and South American countries, with the highest disease burden concentrated in sub-Saharan Africa. However, recent large-scale outbreaks, notably in Brazil from 2016 to 2018, demonstrated a concerning reemergence of YF in regions with historically low incidence. This shift reflects dynamic changes in YFV ecology, including potential viral evolution and vector adaptation, and underscores the vulnerability of previously unaffected or less affected areas [4, 18].

As of May 2025, the Pan American Health Organization (PAHO) reported 235 confirmed YF cases in the Americas, including 96 deaths [5]. This marks an increase from previous years, with cases spreading beyond the Amazon basin into regions such as São Paulo in Brazil and Tolima in Colombia [19, 20]. Similarly, in Africa, the World Health Organization (WHO) has highlighted the need for enhanced surveillance and vaccination efforts to combat the disease [21].

A particularly concerning feature of recent epidemics has been the expansion of YFV into densely populated urban and peri-urban areas. This phenomenon creates a potential bridge for urban transmission, as the highly anthropophilic *Aedes aegypti* mosquito, a primary vector in urban settings, is prevalent in these areas. The convergence of sylvatic and urban ecologies, driven by factors such as deforestation, human encroachment into forested areas, and increased population mobility, significantly elevates the risk of large-scale outbreaks. This complex epidemiological landscape underscores the urgent need for integrated vector surveillance, robust vaccine strategies, and public health preparedness to prevent future epidemics and control the spread of YF [4, 16, 22]. The movement of unvaccinated human populations into or through areas with active sylvatic cycles also contributes to the risk of spillover events and subsequent urban spread.

## 5.2 Challenges in surveillance and reporting

Despite global efforts, the true burden of YF remains challenging to ascertain due to significant limitations in surveillance and reporting systems, particularly in endemic regions. Subclinical infections, where individuals experience no or very mild symptoms, often go undetected, contributing to a substantial underestimation of case numbers. Furthermore, the nonspecific nature of early symptoms can lead to misdiagnosis, with YF often confused with other febrile illnesses prevalent in the same geographical areas, such as dengue, malaria, leptospirosis, or viral hepatitis [4, 10]. This diagnostic ambiguity further complicates accurate reporting.

In recent years, advancements in diagnostic capacity, coupled with the application of molecular and geospatial tools, have significantly enhanced outbreak detection and response capabilities. These tools allow for more precise identification of the virus, tracking of its spread, and mapping of high-risk areas. Nevertheless, limited access to advanced diagnostic facilities in remote or resource-poor areas remains a persistent obstacle to comprehensive surveillance. The lack of robust laboratory infrastructure, trained personnel, and reliable supply chains for reagents hinders timely and accurate diagnosis, thereby impeding effective public health interventions. Improving these aspects is crucial for achieving a more accurate understanding of the disease's epidemiology and for implementing targeted control measures [23].

### 5.3 Conservation medicine and ecosystem dynamics

Understanding YF through the lens of conservation medicine is essential for addressing the complex interactions between environmental changes, wildlife health, and human disease risk [24]. Conservation medicine—closely linked to the One Health framework—emphasizes that the health of humans, animals, and ecosystems is inseparably connected [14].

In Brazil, the sylvatic cycle of YF vividly demonstrates this interdependence. NHPs, especially howler monkeys (*Alouatta* spp.), are highly susceptible to YFV and act as early sentinels for viral circulation in forested areas [17]. Epizootics among these primate populations frequently precede human outbreaks by weeks or months, providing a critical early-warning system [14]. For example, during the 2016–2019 outbreaks, large die-offs of *Alouatta guariba* were detected in the Atlantic Forest biome of Minas Gerais and Espírito Santo, triggering targeted vaccination campaigns in surrounding communities [17].

Ecological pressures such as deforestation, forest fragmentation, and urban expansion have forced NHPs into closer contact with human settlements and peri-urban forest fragments, intensifying the risk of zoonotic spillover [24]. At the same time, changes in vector distribution, driven by habitat alteration and climate variability, have expanded the range of sylvatic mosquitoes like *Haemagogus* and *Sabethes spp.* into areas near urban centers [14].

In practice, Brazil has integrated conservation medicine concepts by combining primate surveillance, entomological monitoring, and remote sensing of forest cover [17]. Health authorities and environmental agencies collaborate to report and investigate sudden primate deaths, collect samples for laboratory testing, and map epizootic clusters to guide emergency vaccination corridors [24]. These integrated actions have proven effective in containing outbreaks and preventing urban transmission [17].

By reinforcing this approach, conservation medicine bridges veterinary science, wildlife ecology, epidemiology, and public health, creating a more robust surveillance network capable of anticipating disease emergence [14, 24]. Sustained investment in training, local community engagement, and intersectoral data sharing are key to maintaining this early-warning capacity and minimizing the impacts of future YF outbreaks [17].

## 6. Clinical presentation

The YFV replicates in various cell types, including monocytes, macrophages, dendritic cells, and hepatocytes. Hepatic involvement is a hallmark of severe YF, leading to the characteristic jaundice that gives the disease its name [9].

The incubation period for YF typically ranges from 3 to 6 days after the mosquito bite, though it can extend up to 10 days. YF presents with a wide clinical spectrum, ranging from asymptomatic or mild infection to severe, life-threatening illness (**Table 1**) [2, 4, 6, 9].

Yellow fever classically presents in three phases. The acute phase begins suddenly with symptoms such as fever, chills, severe headache, backache, generalized myalgia,

| | Infection Phase 3-6 days | Remission Phase 2-24 hours | Intoxication Phase 3-8 days | Convalescence Phase 2 weeks – several months |
|---|---|---|---|---|
| Signs and Symptoms | Fever<br>Headache<br>Asthenia<br>Myalgia<br>Malaise<br>Vertigo<br>Nausea<br><br>Conjunctival infection<br>Bradycardia (Faget sign)<br>White tongue | Alleviation of symptoms<br><br>Phase often goes unnoticed | Headache<br>Epigastric pain<br>Vomiting<br>Myalgia<br>Malaise<br>Jaundice<br>Oliguria → Anuria<br>"Soft liver"<br>Hypotension → Shock<br>Torpor → Coma<br>Hypothermia<br>Hemorrhage<br>Convulsions<br>Encephalopathy<br>Pancreatitis | Some cases might present: Prolonged asthenia, persistent alopecia, progressive jaundice – may occur even if it did not appear during the first 14 days of illness<br><br>Benign outcome: Transaminases and bilirubins return to normal after 1 month without the need for treatment |
| Laboratory Features | Leukopenia<br>Neutropenia<br>C-reactive protein – low<br>↑ AST > ALT<br>Proteinuria<br>↑ LDH | | Thrombocytopenia<br>Leukocytosis<br>↑ AST > ALT<br>Proteinuria<br>Azotemia<br>Hypoglycemia<br>Acidosis<br><br>Recovery or Death | Frequent fluctuations in transaminases occur, and normalization may take months in some patients<br><br>Late Hepatitis – New elevation of transaminases during the convalescent period in 16% of patients |
| Detection | Viremia | | Antibodies | Antibodies |

*Source: Adapted from Brazilian Ministry of Health, 2020.*
*AST: Aspartate Aminotransferase, ALT: Alanine aminotransferase, LDH: Lactate Dehydrogenase.*

**Table 1.**
*Phases of yellow fever virus infection and the main clinical and laboratory characteristics of the disease clinical course of yellow fever. Classically, four periods of the disease are described: infection, remission, intoxication, and convalescence. The symptoms and clinical signs characteristic of each period differ, with a general worsening of the clinical condition observed during the intoxication period, when some patients progress to death. Among those who recover from the disease, around 16% experience a new elevation in transaminases, characterizing a condition known as late-relapsing hepatitis.*

prostration, nausea, and vomiting. This phase typically lasts 3–4 days, after which most patients recover, developing lifelong immunity (**Table 1**) [4, 9, 25].

However, in approximately 15–25% of cases, after a brief remission of symptoms—called a "period of remission" lasting hours to a day—the patient progresses to the more severe toxic phase. This phase is characterized by the reappearance of fever, often accompanied by more severe and systemic manifestations. These include a hemorrhagic diathesis presenting as epistaxis, gingival bleeding, hematemesis (often referred to as "black vomit" due to gastrointestinal hemorrhage), melena, petechiae, purpura, and ecchymoses. Jaundice is indicative of severe liver damage and is the origin of the "yellow" in yellow fever. Renal dysfunction manifests as oliguria, anuria, and acute kidney injury. In severe cases, the disease can lead to multiorgan failure, involving the liver, kidneys, and heart. Neurological symptoms such as convulsions, delirium, or coma may also occur. Mortality in the toxic phase is significant, ranging from 20 to 50%, and can be even higher during epidemics. Death usually occurs within 7 to 10 days of symptom onset [4, 9, 25]. Recent studies have further indicated that elevated levels of aspartate aminotransferase (AST), international normalized ratio (INR), lactate, and WBC, along with older age, were significantly associated with a higher risk of death prior to discharge, highlighting the importance of these clinical markers for improving patient management and the early identification of risk factors and intervention for severe cases (**Table 1**) [25, 26].

More recently, particularly during recent YF outbreaks in Brazil, a phenomenon known as Late Relapsing Hepatitis after Yellow Fever (LHep-YF) has been described in a subset of survivors. This condition is characterized by a rebound in liver enzymes such as AST, alanine aminotransferase (ALT), alkaline phosphatase, and total bilirubin, often occurring within 2 to 6 months after an initial improvement or even normalization of liver function following the acute phase of YF [18, 27–29]. Patients experiencing LHep-YF may present with persistent fatigue, a return of jaundice, headache, hyperoxia, and sometimes low platelet levels. While the exact pathophysiology is not yet fully understood, current hypotheses include viral persistence in the liver, with YFV RNA and antigens having been detected in liver biopsies months after acute infection in some cases [27, 29], or a sustained proinflammatory immune response to the virus [18]. This late complication underscores the importance of extended patient follow-up after acute YF, as it can lead to hospital readmission, though generally, patients do recover, albeit with elevated liver enzymes and symptoms persisting for several months (**Table 1**).

## 7. Pathogenesis

YFV infection leads to a broad spectrum of clinical manifestations, ranging from subclinical or mild febrile illness to fulminant hepatitis with multiorgan failure. The pathogenesis of YF is shaped by a complex interplay between viral replication, host immune responses, and tissue-specific damage—particularly in the liver, spleen, kidneys, and heart (**Figure 3**) [2].

### 7.1 Initial viral replication and dissemination

After inoculation through the bite of an infected mosquito, YFV initially replicates in dendritic cells and macrophages at the skin and draining lymph nodes. This is followed by primary viremia, during which the virus disseminates to major organs, notably the liver, which serves as the primary site of replication (**Figure 3**) [2].

### 7.2 Hepatic involvement

YFV is strongly hepatotropic, preferentially infecting midzonal hepatocytes. Histopathological findings include councilman bodies (eosinophilic apoptotic hepatocytes), steatosis, and necrosis without significant inflammation. This immune-silent hepatic apoptosis results in impaired detoxification and synthesis functions, leading to jaundice, coagulopathy, and elevated transaminases [10, 15]. Even though the YFV usually causes liver damage during its replication, there is a lack of knowledge to understand if alcohol consumption is a risk factor for severe YF (**Figure 3**).

Importantly, a recent study demonstrated that YFV infection induces a redox imbalance in hepatocyte cultures, marked by increased reactive oxygen species (ROS)

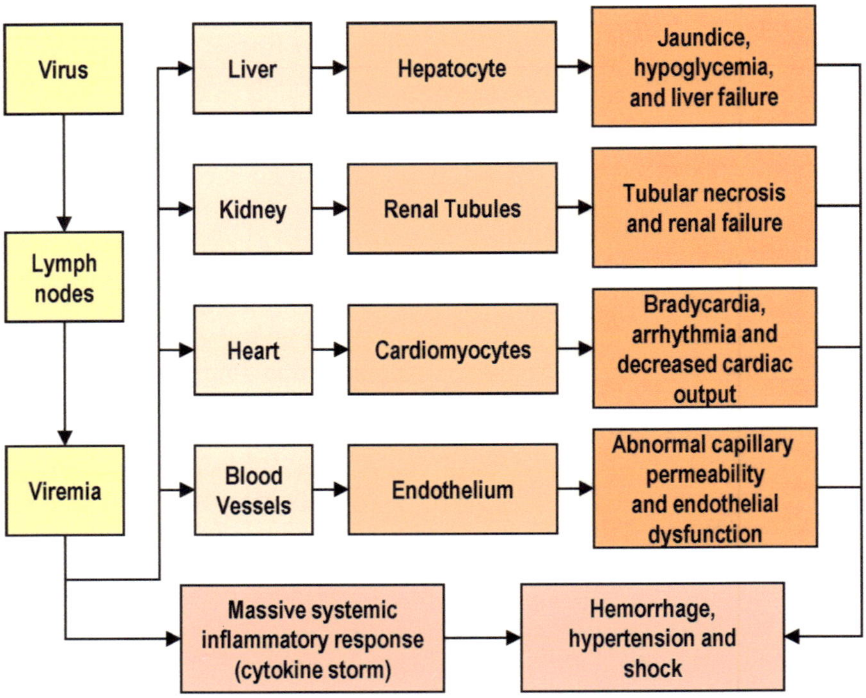


**Figure 3.**
*Pathophysiology of yellow fever. Following inoculation of the yellow fever virus (YFV) into the epidermis of patients by the mosquito vector, the virus migrates to the lymph nodes and then spreads through the bloodstream. Subsequently, the virus reaches systemic organs—primarily the liver, kidneys, heart, and blood vessels—directly and indirectly affecting the tissues of these organs. This leads to the signs, symptoms, and laboratory abnormalities commonly observed in patients with severe yellow fever. In addition to organ involvement, an exacerbated systemic inflammatory response also occurs, contributing to increased capillary permeability, plasma leakage, hypotension, and shock. Source: Adapted from Brazilian Ministry of Health, 2020.*

production and downregulation of antioxidant defenses. This oxidative stress likely contributes to hepatocellular injury and may play a role in systemic manifestations such as shock and multiorgan dysfunction (**Figure 3**) [10].

### 7.3 Immune dysregulation and cytokine storm

Severe YF cases are characterized by a dysregulated immune response, including excessive proinflammatory cytokine production (TNF-α, IL-6, IFN-γ), complement activation, and lymphocyte apoptosis. This "cytokine storm" contributes to vascular leakage, shock, and organ failure. In parallel, YFV has developed mechanisms to evade innate immunity, including inhibition of interferon signaling, which allows uncontrolled viral replication in early stages of infection (**Figure 3**) [30, 31].

### 7.4 Systemic effects and multiorgan failure

In severe cases of YF, systemic viral dissemination leads to widespread tissue damage and multiorgan failure. Renal involvement is particularly common, with acute tubular necrosis and acute kidney injury frequently observed in fatal outcomes. The heart may also be affected, with evidence of myocardial inflammation, arrhythmias, and cardiac dysfunction contributing to hypotension and shock. Although less frequently, the central nervous system (CNS) can be involved, especially in severe or fulminant cases. Histopathological and immunohistochemical analyses have confirmed viral neuroinvasion in fatal human cases, indicating that YFV can cross the blood-brain barrier under certain pathological conditions (**Figure 3**) [30].

## 8. Host immune response

The study of the immune response is of great relevance for understanding the evolution of tissue lesions inherent to different infections. Specifically, in viral infections that affect the liver, such as hepatitis B, C, dengue, and YF, cytotoxic T cell responses and cytokine activity are important factors involved in controlling viral replication and hepatocyte death [32].

A comparative study examining how the wild-type YF virus Asibi strains and the vaccine strain (17D) infect host cells revealed that mutations in the E protein of the vaccine strain 17D acquired during the attenuation process result in differentiation of the infection pathway. It was observed that the Asibi strain infects target cells exclusively *via* classical clathrin-mediated endocytosis, whereas the 17D strain utilizes a clathrin-independent pathway for infection. Interestingly, this differentiation led the vaccine strain to bind and infect host cells more efficiently than the wild-type Asibi strain (**Figure 4**). Consequently, these characteristics culminated in enhanced viral replication and greater release of viral RNA into the host cell cytosol, which in turn may activate pattern recognition receptors (PRRs) and lead to a more robust cytokine-mediated antiviral response, notably interferon (IFN) [33].

Another immune response to viral infection that occurred more robustly in the presence of the 17D strain was the secretion of chemokines CXCL10 and CCL5 [33], which are almost invariably associated with viral infections [34] and primarily function by attracting leukocytes to sites of viral infection (**Figure 4**). These findings suggest that entry mechanisms are responsible for most of the differences in cytokine responses observed in cells infected with both strains. However, the attenuation of 17D is likely a multifactorial process that depends on additional mechanisms [33].

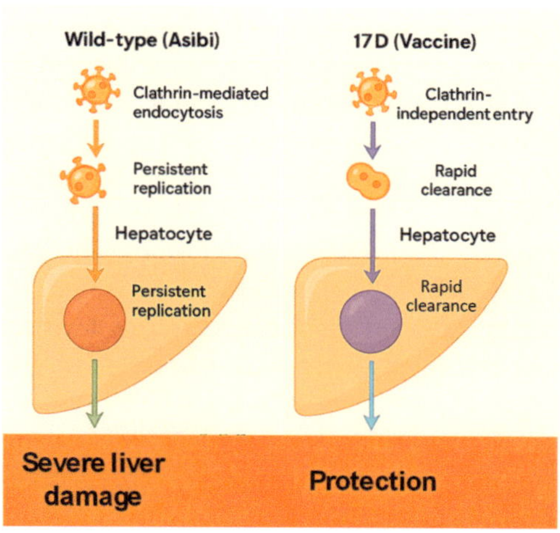


**Figure 4.**
*Comparison of host immune responses to wild-type yellow fever virus (Asibi) and the 17D vaccine: Clathrin-independent entry.*

In general, therefore, enhanced replication in the early phase of infection, possibly due to attenuation of the 17D strain, leads to rapid viral elimination. In contrast, the wild-type strain, infecting cells less efficiently, may replicate at lower levels within cells, passively evading detection by innate immune components. This would allow the virus to spread and reach its target organs (**Figure 4**) [33].

Surviving YF infection requires the host to generate a controlled immune response that recognizes and eliminates the invading pathogen while limiting collateral damage to its own tissues, which may result from an exaggerated immune response [28, 35–37]. Immunoregulation may emerge as a result of the host's response to infection to maintain or restore a homeostatic environment and may be actively induced by the pathogen to ensure its survival [36].

Previous studies investigating the behavior of the wild-type YF virus and the live-attenuated YF 17D strain in human hepatocytes and Kupffer cells have shown distinct responses. Infection with YF 17D led to limited viral spread and an immune response indicative of rapid viral elimination and immunological protection. On the other hand, an expressive response to wild-type YF infection, suggesting a "cytokine storm," indicated a critical component in disease progression (**Figure 4**) [38].

A more detailed assessment of cytokine and chemokine production by human Kupffer cells infected with wild-type or vaccine strain YFV revealed that both strains support viral replication and induce a prominent and prolonged proinflammatory response characterized by elevated levels of TNF-α, CXCL8, and CCL5, with a lesser increase in IL-10. Notably, cytokine and chemokine levels throughout the infection were higher in cells infected with the wild-type virus compared to those infected with the vaccine strain. These findings suggest that differentially regulated infection in Kupffer cells may play a critical role in disease development [39].

Furthermore, Cong et al. [36] evaluated the susceptibility of macrophages and dendritic cells to YFV infection, the cytokine profile induced by the infection, and their ability to interact with and stimulate $CD4^+$ T cells. Their data demonstrated that both human and nonhuman primate-derived macrophages and dendritic cells were susceptible to infection with both wild-type (Asibi) and vaccine (17D) strains. Both stimulated distinct cytokine responses, while dendritic cells infected with the vaccine virus were able to stimulate $CD4^+$ T cells, whereas cells infected with the wild-type virus did not. These findings suggest that the wild-type YFV may inhibit critical components of the immune response to allow viral dissemination and spread. Indeed, IFN-α and TNF-α levels were elevated 2–5 days after infection with the vaccine strain, with a decrease in response as the infection progressed. In contrast, cells infected with the wild-type virus maintained a persistent and increased response as the infection progressed. Analysis of CCL2, CCL3, and CCL5 expression also suggested a controlled response to infection with the vaccine strain, with high expression of these chemokines 2–5 days after infection, while their expression in cells infected with the wild-type virus was delayed and significantly elevated later [36].

A study using a hamster model of YFV infection was developed to assess the dynamics of cytokine expression changes in liver tissue during infection. It was observed that levels of proinflammatory cytokines (IFN-γ, IL-2, and TNF-α) were reduced 8 days after infection and increased by day 12. Regarding modulating cytokines, IL-10 expression was increased, and TGF-β expression was decreased in the liver of infected animals [40]. From a pathological standpoint, the observed cytokine profile suggests that YFV infection has immunosuppressive effects, which would contribute to liver damage in the intermediate stages of infection, followed by pathogenic mechanisms leading to disease progression in the later stages.

Similarly, a study conducted in IFN system deficient murine models, such as mice deficient in receptors for IFN-α/β (A129) or IFN-γ (G129), both IFN-α/β and IFN-γ receptors (AG129), or the signaling molecule STAT1 (STAT129), demonstrated severe liver and spleen disease with very high levels of CCL2 and IL-6, suggestive of a cytokine storm, associated with rapid viral dissemination and extensive replication in these visceral organs. For the 17D-204 vaccine strain, it was observed that infection of visceral organs was rarely established and rapidly eliminated, possibly through IFN-γ-dependent mechanisms [41].

A pioneering approach by ter Meulen and colleagues (2004) evaluated systemic mediators (IL-6, TNF-α, CXCL8, CCL2, CXCL10, and IL-1Ra) in the pathogenesis of YF. The levels of soluble mediators were compared in patients with fatal YF versus those with nonfatal YF. Elevated levels of IL-6, CCL2, CXCL10, and IL-1Ra were verified in patients with fatal YF compared to those with nonfatal YF. These distinct patterns of inflammatory and regulatory biomarkers associated with fatal or nonfatal YF paralleled viremia, suggesting a significant contribution of these mediators to the terminal complications of the disease [35].

In addition, Quaresma et al., by using immunohistochemistry to characterize the phenotype of inflammatory cells and assess the presence of apoptosis in the liver, demonstrated a predominance of $CD4^+$ T lymphocytes, accompanied by $CD8^+$ T lymphocytes, natural killer cells, macrophages, and antigen-presenting cells [32]. Murai et al. demonstrated that the migration of activated $CD8^+$ T cells from lymph nodes to the liver is mediated by the CCR5-CCL5 axis [42]. However, the exact function of each of these specific cell populations in the immunopathogenesis of YF remains unknown. Furthermore, the authors observed a disproportion between the intensity of inflammation and the degree of liver damage, which is likely associated with intense apoptotic activity, which classically does not induce a significant inflammatory response [32]. This finding is a hallmark of the liver in this disease, previously observed in classic studies by Hudson [43], Klotz and Belt [44, 45], and later confirmed by Smetana [46], Vieira et al. [47], Branquet [48], and Quaresma et al. [49], both in human and experimental material, evaluating infected nonhuman primates. Despite this disproportion, the cellular immune response plays an important role in the pathogenesis of hepatocellular injury observed in severe YF, likely as a consequence of cytolytic actions involving class II MHC and the activation of Fas receptors and granzymes/perforins [32].

In YF patients who progressed to death, the histopathological picture is characterized by lytic steatosis, necrosis, and apoptosis associated with the presence of moderate mononuclear inflammatory infiltrate. The inflammatory infiltrate mainly consisted of $CD4^+$ T lymphocytes, followed by $CD8^+$ T lymphocytes, with immunoreactivity for the Fas ligand, indicating the importance of lymphocytes in establishing liver lesions in severe YF, inducing hepatocyte apoptosis through binding to Fas receptors [50].

Necrosis is likely a consequence of the direct cytopathic effect of YFV, associated with a potent immune response in which $CD4^+$ and $CD8^+$ T lymphocytes and cytokines, especially TGF-β, but also TNF-α and IFN-γ, play an important role [51]. These data suggest that patients who progress to the severe form of the disease exhibit a T cell response that may be ineffective in controlling viral replication in the liver or is strongly cytolytic, causing damage and exacerbating pathology [52].

Still analyzing patients who progressed to death, Quaresma et al. [53] demonstrated that, in paraffin-embedded liver blocks, the subpopulations of infiltrating cells at the sites of inflammation are mainly composed of $CD4^+$ T lymphocytes, with a small number of $CD8^+$ cytotoxic lymphocytes, $CD20^+$ B lymphocytes, $NKT^+$ cells, and dendritic cells. Some cells expressed TNF-α and IFN-γ, but a much larger proportion of cells expressed TGF-β, suggesting a Th1 and Th3 immune response pattern, respectively. Viral antigens, whose production could interfere with hepatocyte biology, would induce the activation of the apoptosis cascade, but TGF-β was also an apoptosis promoter. The most affected hepatocytes presented apoptosis markers that appear in the main cell death pathway in this infection [53].

In a prospective study by Fradico et al. [31], hospitalized patients with confirmed YF were analyzed for circulating chemokines, cytokines, and growth factors. The study presented a detailed immunological landscape, highlighting that distinct immune response patterns underlie disease pathogenesis, progression, and the development of late-relapsing hepatitis in YF patients. A storm of soluble mediators was observed in the acute phase of YF, with elevated levels of inflammatory mediators associated with higher morbidity scores, the need for intensive care, and fatal outcomes. Patients who did not develop hepatitis exhibited a peak in biomarker levels during the acute phase. Conversely, patients who developed late-relapsing hepatitis showed a peak in biomarker levels during the convalescent phase [31].

Taken together, these immune signatures suggest that an imbalance between proinflammatory drive and regulatory mechanisms contributes to immune-mediated tissue damage. Several of these soluble mediators have been proposed as early prognostic biomarkers, potentially useful for risk stratification and clinical screening during outbreaks. Furthermore, their association with late-relapsing hepatitis, a delayed complication marked by recurrent hepatic inflammation, highlights their usefulness in patient follow-up. Incorporating immune profiling into routine care could aid in identifying high-risk individuals and tailoring supportive interventions [26, 31, 54].

## 9. Laboratory findings and diagnosis

Diagnosing YF requires a combination of clinical suspicion, epidemiological context, and laboratory confirmation.

### 9.1 Laboratory findings

Regarding hematologic findings, leukopenia, especially lymphopenia, is common in the acute phase, and thrombocytopenia and neutrophilia is usually present in severe cases of YF, contributing to hemorrhagic manifestations. Anemia may develop due to bleeding [4, 9, 25]. Hepatic indicators include elevated liver enzymes (AST, ALT), which are characteristic and often reach very high levels in severe disease. Hyperbilirubinemia (both conjugated and unconjugated) leads to jaundice. Prolonged prothrombin time (PT) and international normalized ratio (INR) reflect impaired hepatic synthesis of clotting factors, and hypoglycemia may occur due to liver failure [9]. For renal function, elevated creatinine and urea nitrogen indicate kidney involvement, and proteinuria is common. Finally, electrolyte imbalances may be present, particularly in cases with vomiting or renal dysfunction [9, 25, 55].

### 9.2 Diagnosis

Laboratory confirmation is crucial, especially given the nonspecific nature of early symptoms and the need to differentiate from other febrile illnesses such as malaria, dengue, leptospirosis, or viral hepatitis [56]. Specific laboratory diagnosis of YF can

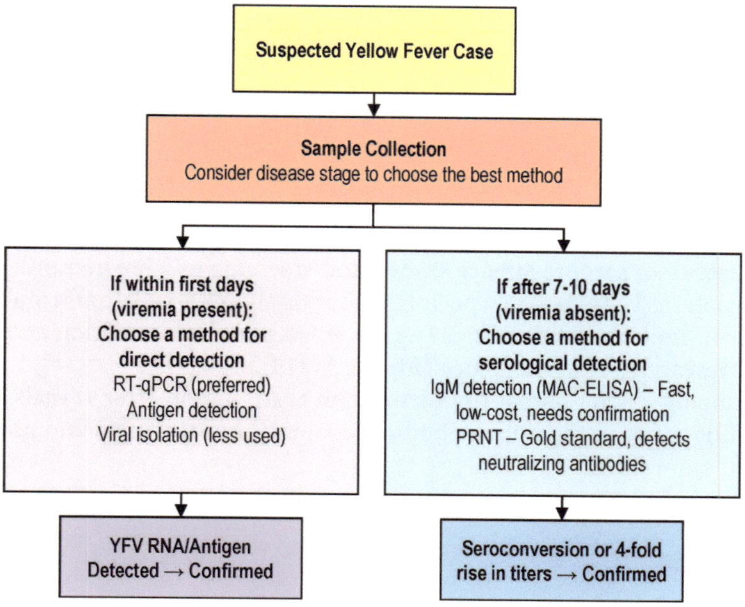


**Figure 5.**
*Diagnostic methods for yellow fever according to disease stage. Flowchart illustrating recommended diagnostic approaches for yellow fever based on the timing of sample collection relative to symptom onset. In the early phase of infection (day 6-7), when viremia is typically present, direct detection methods such as RT-qPCR (preferred), antigen detection, and viral isolation can be employed. In later stages (after day 7), when viremia has usually subsided, serological methods are indicated. These include MAC-ELISA for IgM detection and Plaque Reduction Neutralization Test (PRNT), which is the gold standard for confirming yellow fever diagnosis by detecting neutralizing antibodies.*

be performed using various methods for direct or indirect detection of the virus. The disease stage at sample collection must be considered when choosing the most appropriate diagnostic method (**Figure 5**) [1, 6, 55].

In the early stages, while viremia is present, direct detection methods can be used, such as antigen detection tests, viral isolation in cell culture, or viral RNA detection by molecular techniques (**Figure 5**).

YFV detection using reverse-transcriptase real-time polymerase chain reaction (RT-PCR) (**Figure 5**). This is the preferred method for early diagnosis due to higher sensitivity and specificity and faster, conclusive results compared to methods like viral isolation. This technique detects viral RNA, especially within the first 5–7 days of symptom onset when viremia is highest. It can be performed on blood, urine, or tissue samples, including liver biopsies or autopsy samples, and is useful for distinguishing between wild-type infection and vaccine-associated adverse events [2, 6, 13, 56–58].

Serological tests: Since viremia is often short in YFV infections, serological tests may be needed to confirm the YF diagnosis (**Figure 5**). Assays such as IgM ELISA (MAC-ELISA) detect YFV-specific IgM antibodies, which typically appear 10–15 days after symptom onset and persist for several months, indicating recent infection. MAC-ELISA is widely used for YF diagnosis because it is less expensive, less laborious, and faster. However, because cross-reactivity with other flaviviruses like dengue and Zika can occur, confirmatory tests may be required. Noteworthy, neutralization assays, such as the Plaque Reduction Neutralization Test (PRNT), are considered the gold standard method for serological diagnosis of YF and the best correlate of protection. PRNT, despite being laborious and time-consuming (requiring about 7 days for results), offers high sensitivity, specificity, and accuracy with the ability to differentiate YFV infection from other flavivirus infections by measuring antibody titers [1, 13, 57, 59, 60]. PRNT is particularly useful for confirming ambiguous IgM-positive results [13]. Indeed, in outbreak and epidemic scenarios, accurate detection of neutralizing antibodies is essential, aiding both surveillance and evaluation of vaccine efficacy and coverage [61]. Additionally, IgG ELISA detects YFV-specific IgG antibodies, which develop later than IgM and indicate past infection or vaccination; this test is useful for seroprevalence studies and assessing vaccine immunity [13].

Virus isolation: Though rarely performed routinely, YFV isolation from blood (during the viremic phase) or tissues (e.g., liver) can confirm infection and is primarily used for research and surveillance (**Figure 5**) [13].

Histopathology: Liver biopsy or postmortem examination often reveals characteristic findings such as councilman bodies (apoptotic hepatocytes) and midzonal necrosis [9].

## 10. Management

Since there is no specific antiviral treatment for YF, patient management is entirely supportive. The goal is to alleviate symptoms, prevent complications, and maintain vital organ function. Early recognition and intensive supportive care are critical for improving outcomes in severe cases [30, 62].

Key aspects of management include fluid and electrolyte administration, where intravenous fluids are essential to maintain hydration, blood pressure, and support renal function, particularly in patients with vomiting, diarrhea, or signs of shock. Management of hemorrhage involves close monitoring for bleeding. Blood transfusions, including packed red blood cells, fresh frozen plasma, and platelets, may be required for substantial hemorrhage or severe coagulopathy. For renal support, strict input and output monitoring is crucial, and dialysis may be necessary for patients who develop acute kidney injury with oliguria/anuria or severe electrolyte disturbances. Close monitoring of liver function tests and coagulation parameters, along with managing hypoglycemia if present, is necessary. Since there is no specific treatment to reverse liver damage, care focuses on supporting the patient until liver regeneration occurs [30, 62].

Finally, patient isolation is important. Individuals with YF should be protected from mosquito bites using mosquito nets for at least 5 days after symptom onset to prevent further transmission to uninfected mosquitoes and, subsequently, to other humans [62].

## 11. Treatment, control measures, and prevention

Currently, there are no specific antiviral treatments available against YFV with proven efficacy. Some antivirals used for other diseases, such as ribavirin (used for canine distemper and other important human viral infections) and sofosbuvir (commonly used against hepatitis C infection), have been tested *in vitro* and *in vivo* against YFV. However, studies with these drugs have been inconclusive or failed to demonstrate efficacy in prolonging the survival of YFV-infected individuals or animals. New molecules have also been tested against YFV *in vitro*, and some have shown antiviral activity against this virus and other flaviviruses, but none are currently available for clinical use [2, 57]. The absence of specific drugs to treat this disease limits its management, meaning that treatment of severe cases relies solely on life-support measures [63]. Both individual and collective control and prevention measures can and should be implemented against YF in humans.

Established strategies for YF control encompass vector surveillance and control programs aimed at monitoring mosquito population densities and species composition in areas at risk of transmission. Additionally, vaccination campaigns are implemented to protect vulnerable populations and individuals at increased risk of exposure due to travel, recreational activities, or occupational engagements. In addition, case investigation and reporting of human cases and epizootics are highly relevant actions for implementing viral containment measures, thereby reducing disease impact and helping prevent outbreaks and epidemics [58, 64, 65]. It is important to note that complete elimination of YFV from nature is impossible, as the sylvatic cycle of the disease does not allow interventions to control or prevent viral circulation and epizootics. However, preventing human cases is possible through measures such as vaccination, regarded as the primary and most important individual protective measure against YF [66, 67].

### 11.1 Yellow fever 17DD vaccine

The 17D and 17DD vaccines were developed in 1936 by a group of researchers at the Rockefeller Foundation in the United States and have been used worldwide with high efficacy since then, preventing millions of severe cases and deaths [67]. These live-attenuated virus vaccines were developed through sequential passages of a wild-type YFV sample—isolated from the blood of a febrile patient in Ghana (called the Asibi strain)—in various cell cultures restrictive to its replication. Over 200 passages were performed in embryonic tissues of mice and chickens, with viruses from each passage being tested in murine and non-human primate (NHP) models to assess neurovirulence and viscerotropism. These tests demonstrated that sequential passaging of the YFV produced an attenuated viral sample that lost the ability to cause fever, tissue damage, hepatitis, and viremia in NHPs [67–69]. The first clinical trials with this attenuated viral sample were carried out in 1936, after about 227 passages in different tissues. From this stock, two samples were derived—named 17D-204 and 17DD—which are still used today in the production of certified commercial YF vaccines.

The differences between the 17D-204 and 17DD vaccine strains are only a few passages in embryonated eggs, which did not cause major genomic changes. Thus, they are considered identical in efficacy and safety. Since their development, the YF vaccines have been produced in embryonated chicken eggs. Currently, four WHO-certified manufacturers, based in Brazil, Senegal, Russia, and France, produce and distribute the 17DD and 17D vaccines, with Brazil currently being the largest producer [67, 69, 70].

The 17D/17DD vaccine is known to induce a fast, specific, potent, and long-lasting immune response similar to wild-type virus infection. This response is primarily mediated by neutralizing antibodies targeting the envelope (E) protein, conferring protection that can last decades or even a lifetime [67, 71, 72]. The YFV vaccine is considered one of the safest and most effective vaccines ever developed and has been the subject of extensive studies aimed at understanding long-term immune memory. Immunization studies have shown that a single dose was able to induce the production of neutralizing antibodies in up to 90% of recipients within 10 days and in nearly all individuals by 30 days postvaccination. Circulating antibody levels generally remain stable for many years, and in some individuals, up to 65 years after vaccination [67, 69, 73]. Based on these findings, WHO currently recommends a single vaccine dose for adults not living in endemic areas. However, individuals who live, work, or frequently travel to high-risk areas are advised to receive a booster dose 10 years after the first dose [74].

Although the immune response profile induced by the vaccine is very similar to that elicited by the wild-type virus, the attenuation process of 17DD is occasioned by several genomic mutations, allowing the biological changes needed for attenuation [66, 67]. Sixty-seven genetic mutations have been identified between the 17D-204 vaccine strain and the Asibi parent strain; twenty-two of these led to amino acid changes. Eight of these amino acid substitutions are in the sequence encoding the envelope (E) protein—the main target of humoral responses—primarily due to the presence of neutralization and hemagglutination epitopes [37, 68]. In addition to its high immunogenic potential, the E protein mediates viral binding, fusion, and adsorption to host cells, playing a fundamental role in viral multiplication and successful infection [71, 72]. Some amino acid substitutions in the vaccine strain result in conformational changes that eliminate certain antibody epitopes present on the wild virus's envelope, meaning these viruses have distinct antigenic profiles recognized by different antibodies [68, 75]. Recent studies also show significant differences in domains I and II of the E protein in South American wild YFV genotypes compared to the 17DD vaccine strain and African wild-type strains. These differences may affect recognition by neutralizing antibodies and could explain the lower immune response in individuals vaccinated with 17DD or 17D when exposed to South American wild-type viruses [75].

Mutations in the 17DD E protein have also changed the virus's cell-entry pathways, enabling use of mechanisms beyond the clathrin-mediated route used by wild-type YFV [33]. This gives the vaccine virus a biological advantage, allowing it to infect host cells more efficiently and quickly. *In vivo*, this advantage is seen as a faster and higher viremia peak in vaccinated individuals, unlike wild-type viruses, which produce lower and more prolonged viremia peaks [31]. Such differences in viral replication profiles may trigger a stronger immune response to the vaccine strain and quicker viral clearance. Wild strains replicate more slowly and at lower levels, which facilitates evasion of some immune mechanisms.

#### *11.1.1 Adverse events following vaccination (AEFV)*

Although the 17D/17DD YF vaccine is widely regarded as safe and highly effective, rare but serious adverse events following vaccination (AEFV) have been documented, particularly in individuals with specific risk factors. These include elderly people over 60 years of age, as well as individuals with thymic disorders, immunodeficiencies, or autoimmune diseases. The most clinically relevant severe AEFVs are categorized into three types:

Yellow Fever Vaccine-Associated Viscerotropic Disease (YEL-AVD) is a rare but life-threatening complication that mimics wild-type YF, with multiorgan involvement including hepatic, renal, and respiratory failure. It occurs at an estimated rate of 0.3 cases per 100,000 vaccine doses and is associated with a high case-fatality rate approaching 60% [4, 76].

Yellow Fever Vaccine-Associated Neurotropic Disease (YEL-AND) encompasses a spectrum of neurological complications such as meningoencephalitis, Guillain-Barré syndrome, and acute disseminated encephalomyelitis (ADEM). It occurs at an estimated rate of 0.4 cases per 100,000 vaccine doses. These events are more frequently reported in infants and older adults [4, 76].

Anaphylaxis, while extremely rare, can occur in individuals with severe egg protein allergy, as the vaccine is propagated in embryonated chicken eggs [4, 76].

#### *11.1.2 Contraindications and special populations*

The YF vaccine is contraindicated in several populations due to the risk of severe adverse effects. These include (i) infants under 6 months of age; (ii) individuals with severe allergies to any component of the vaccine; (iii) persons with active HIV/AIDS infection or other immunosuppressive conditions, except in cases where benefits outweigh risks under controlled clinical settings; and (iv) pregnant women, unless exposure is unavoidable due to outbreak situations or imminent travel to endemic areas [4, 76].

For older adults, especially those above 60 years, a careful risk-benefit assessment is recommended due to the increased incidence of severe AEFIs observed in this group [4, 76].

### 11.2 Challenges and perspectives

Despite the proven efficacy of the 17D/17DD YF vaccine, several challenges continue to hinder YF prevention and control efforts. The reemergence of *Aedes aegypti* in urban settings raises concern about the potential reurbanization of YF transmission, especially in regions with inadequate vector control. Additionally, vaccine hesitancy, fueled by misinformation, has led to suboptimal vaccine coverage in some populations. Moreover, limited laboratory capacity in outbreak zones often results in delayed diagnosis and response [77–79].

To address these issues, next-generation vaccines are under development, including nonreplicating platforms designed to enhance safety in immunocompromised individuals and special populations [77–79]. These advances offer promising avenues to expand vaccination coverage while minimizing risk.

## 12. Conclusion

Yellow fever remains a significant public health threat in endemic regions of Africa and South America, despite being a vaccine-preventable disease. The resurgence of sylvatic YF in Brazil in 2017–2018, the 2024–2025 outbreaks in the Amazon region, and the potential for reurbanization of transmission underscore the need for sustained vigilance and strengthened surveillance systems.

The YFV exemplifies a complex host-pathogen interaction, in which the host immune response plays a dual role. On one hand, it is essential for viral control and recovery; on the other, when dysregulated, the immune response contributes directly to YF pathogenesis and poor outcomes, particularly through the mechanisms of immune hyperactivation, cytokine storm, and immune exhaustion.

Advances in understanding immune biomarkers, as demonstrated by recent clinical studies, have provided valuable tools for prognostication and for guiding patient management. Moreover, the identification of late-relapsing hepatitis as a post-acute manifestation has expanded the clinical spectrum of YF and highlighted the need for longitudinal follow-up of survivors.

The 17D/17DD live-attenuated vaccine remains a cornerstone of prevention with proven efficacy and safety. Nonetheless, rare but serious adverse events call for improved risk assessment strategies, particularly in vulnerable populations such as the elderly and immunocompromised. There is also growing interest in the development of next-generation vaccines with enhanced safety profiles.

In conclusion, while substantial progress has been made in the understanding and control of YF, continuous scientific research, public health investment, and global collaboration are essential to prevent future outbreaks and mitigate the burden of this reemerging disease.

## Acknowledgements

The authors would like to thank the Oswaldo Cruz Foundation (Fiocruz-Minas) for institutional support and infrastructure, as well as the entire yellow fever collaborative group for their invaluable contributions to yellow fever research. Special thanks to the patients and health professionals involved in the surveillance and management of yellow fever in Brazil. This work was partially supported by grants from Brazilian national funding agencies, including CNPq, CAPES, and FAPEMIG.

## Conflict of interest

The authors declare no conflict of interest.

## Glossary

| | |
|---|---|
| ADEM | acute disseminated encephalomyelitis: rare inflammatory condition affecting the central nervous system, associated with neurotropic adverse events after vaccination. |
| AEFV | adverse event following vaccination: any untoward medical occurrence following vaccination, which may or may not be causally related. |
| ALT | alanine aminotransferase: enzyme indicating liver function, elevated in hepatocellular injury. |
| AST | aspartate aminotransferase: another key liver enzyme, often high in yellow fever. |
| $CD4^+$ T cells | helper T lymphocytes: coordinate immune responses by activating B cells and cytotoxic T cells. |
| $CD8^+$ T cells | cytotoxic T Lymphocytes: destroy virus-infected cells through direct cytotoxicity. |
| CFR | case fatality rate: proportion of deaths among confirmed cases. |
| CNS | central nervous system: brain and spinal cord; can be affected in severe yellow fever. |
| CXCL10 | interferon gamma-induced protein 10: a chemokine associated with T cell recruitment and inflammation. |
| FFP | fresh frozen plasma: blood product used to manage coagulopathy. |

| | |
|---|---|
| IgG | immunoglobulin G: main antibody providing long-term protection. |
| IgM | immunoglobulin M: first antibody produced during acute infection. |
| IL-1Ra | interleukin-1 receptor antagonist: a regulatory cytokine that modulates inflammation. |
| IL-6, IL-10, IL-18 | interleukins: cytokines involved in pro- and anti-inflammatory responses. |
| IPT | Incubation period time: time between infection and symptom onset. |
| LRH | late-relapsing hepatitis: delayed recurrence of liver inflammation after apparent recovery. |
| NHP | non-human primates: natural reservoirs of sylvatic yellow fever. |
| PRNT | plaque reduction neutralization test: laboratory test for specific flavivirus antibody confirmation. |
| ROS | reactive oxygen species: molecules that contribute to oxidative stress and tissue damage. |
| RT-PCR | reverse transcription polymerase chain reaction: molecular test for detecting viral RNA. |
| sCD14, sCD163 | soluble CD14/CD163: biomarkers indicating monocyte/macrophage activation. |
| Th1 | type 1 helper T cell: subset of $CD4^+$ T cells producing IFN-$\gamma$ and supporting cell-mediated immunity. |
| Treg | regulatory T cell: subset of T cells that suppress excessive immune responses to prevent tissue damage. |
| WHO | World Health Organization: United Nations agency responsible for international public health. |
| YEL-AVD | yellow fever vaccine-associated viscerotropic disease: severe adverse event mimicking wild-type yellow fever. |
| YEL-AND | yellow fever vaccine-associated neurotropic disease: neurological complications after vaccination. |
| YF | yellow fever: the disease caused by the yellow fever virus. |
| YFV | yellow fever virus: the causative agent of yellow fever. |

**Author details**

Jordana Rodrigues Barbosa Fradico[1*], Izabela Maurício de Rezende[2], Andreza Parreiras Gonçalves[1], Jaquelline Germano De Oliveira[1], Ana Carolina Campi-Azevedo[1], Olindo Assis Martin-Filho[1] and Andréa Teixeira-Carvalho[1]

1 Oswaldo Cruz Foundation (FIOCRUZ), René Rachou Institute, Belo Horizonte, Minas Gerais, Brazil

2 Departament of Pediatric Infectious Diseases, Stanford University School of Medicine, Stanford, CA, United States

*Address all correspondence to: jordanafradico@gmail.com

## References

[1] Staples JE, Monath TP. Yellow fever: 100 years of discovery. Journal of the American Medical Association. 2008;**300**(8):960-962. DOI: 10.1001/jama.300.8.960

[2] Monath TP, Vasconcelos PFC. Yellow fever. Journal of Clinical Virology. 2015;**64**:160-173. DOI: 10.1016/j.jcv.2014.08.030

[3] Barrett AD. Yellow fever in Angola and beyond--the problem of vaccine supply and demand. The New England Journal of Medicine. 2016;**375**(4):301-303. DOI: 10.1056/NEJMp1606997

[4] Silva NIO et al. Recent sylvatic yellow fever virus transmission in Brazil: The news from an old disease. Virology Journal. 2020;**17**:9. DOI: 10.1186/s12985-019-1277-7

[5] PAHO. PAHO - Epidemiological Alert Yellow Fever in the Americas Region 31 May 2025. 2025. Available from: https://www.paho.org/en/documents/epidemiological-alert-yellow-fever-americas-region-31-may-2025 [Accessed: July 10, 2025]

[6] WHO. WHO - Yellow Fever. 2022. Available from: https://www.who.int/emergencies/emergency-events/item/yellow-fever [Accessed: April 16, 2025]

[7] Pierson TC, Diamond MS. Flaviviruses. In: Fields BN, Knipe DM, Howley PM, editors. Fields Virology. Philadelphia: Lippincott Williams & Wilkins; 2013. pp. 747-794

[8] Lindenbach BD, Rice CM. Molecular biology of flaviviruses. Advances in Virus Research. 2003;**59**:23-61. DOI: 10.1016/s0065-3527(03)59002-9

[9] Monath TP. Yellow fever: An update. The Lancet Infectious Diseases. 2001;**1**(1):11-20. DOI: 10.1016/S1473-3099(01)00016-0

[10] Coelho Ferraz A et al. Yellow fever virus infection in human hepatocyte cells triggers an imbalance in redox homeostasis with increased reactive oxygen species production, oxidative stress, and decreased antioxidant enzymes. Free Radical Biology and Medicine. 2024;**213**:266-273. DOI: 10.1016/j.freeradbiomed.2024.01.042

[11] ICTV. ICTV - Book: Flaviviridae. 2025. Available from https://ictv.global/report/chapter/flaviviridae/flaviviridae/orthoflavivirus [Accessed: April 18, 2025]

[12] Barrett ADT, Monath TP. Epidemiology and Ecology of YELLOW fever Virus. 2003;**61**:291-315. DOI: 10.1016/S0065-3527(03)61007-9

[13] Staples JE, O'Laughlin K. Yellow fever. Yellow book 2026: health information for international travel. In: Halsey E, editor. Center for Disease Control and Prevention (CDC). Oxford: Oxford University Press; 2025

[14] Vasconcelos PFC. Yellow fever in Brazil: Thoughts and hypotheses on the emergence in previously free areas. The Revista de Saúde Pública. 2010;**44**(6):1144-1149. DOI: 10.1590/S0034-89102010005000046

[15] Dantas Andrade VDG et al. Re-emergence of yellow fever virus in Brazil: Evidence from forest and peri-urban settings. medRxiv [Preprint]. 2025:25325467. DOI: 10.1101/2025.04.10.25325467

[16] Faria NR et al. Genomic and epidemiological monitoring of yellow fever virus transmission potential. Science. 2018;**361**(6405):894-899. DOI: 10.1126/science.aat7115

[17] Silva NIO et al. Ecological drivers of sustained enzootic yellow fever virus transmission in Brazil, 2017-2021. PLos – Neglected Tropical Disease Peer-Reviewed. 2023;**7**(6):e0011407. DOI: 10.1371/journal.pntd.0011407

[18] Rezende IM et al. Characterization and investigation of risk factors for late-relapsing hepatitis after yellow fever. Clinical Infectious Diseases. 2023;**77**(4):565-573. DOI: 10.1093/cid/ciad249

[19] PAHO. PAHO Issues Epidemiological Alert Due to Increase in Yellow Fever Cases in Latin America. 2025. Available from: https://www.reddit.com/r/ContagionCuriosity/comments/1ihutv6/paho_issues_epidemiological_alert_due_to_increase/?utm_source=chatgpt.com [Accessed: April 21, 2025]

[20] PAHO. Epidemiological Alert: Increase in Yellow Fever Cases in Latin America. 2025. Available from: https://www.paho.org/en/news/4-2-2025-epidemiological-alert-increase-yellow-fever-cases-latin-america?utm_source=chatgpt.com [Accessed: April 20, 2025]

[21] WHO. Yellow Fever – African Region (AFRO). 2024. Available from: https://www.who.int/emergencies/disease-outbreak-news/item/2024-DON510?utm_source=chatgpt.com [Accessed: April 23, 2025]

[22] Xavier AR et al. Yellow fever: Laboratorial diagnosis and clinical manifestations. Jornal Brasileiro de Patologia e Medicina Laboratorial. 2018;**54**(5):296-305. DOI: 10.5935/1676-2444.20180050

[23] World Health Organization. A Global Strategy to Eliminate Yellow Fever Epidemics 2017-2026 [Internet]. Geneva: World Health Organization (WHO); 2018. Licence: CC BY-NCSA 3.0 IGO. Available from: https://iris.who.int/bitstream/handle/10665/272408/9789241513661-eng.pdf?sequence=1&isAllowed=y [Accessed: April 26, 2025]

[24] Saad LDC, Chiaravalloti-Neto F. Reemergence of yellow fever in the state of São Paulo: The structuring role of surveillance of epizootics in non-human primates in a one health approach. Revista Brasileira de Epidemiologia. 2024;**27**:e240064. DOI: 10.1590/1980-549720240064

[25] McClure M et al. Risk factors associated with in-hospital mortality during yellow fever outbreak in Brazil. Frontiers in Medicine (Lausanne). 2025;**12**:1505005. DOI: 10.3389/fmed.2025.1505005

[26] Kallas EG et al. Predictors of mortality in patients with yellow fever: An observational cohort study. The Lancet Infectious Diseases. 2019;**19**:750-758. DOI: 10.1016/S1473-3099(19)30125-2

[27] Rezende IM et al. Late-relapsing hepatitis after yellow fever. Viruses. 2020;**12**(2):222. DOI: 10.3390/v12020222

[28] Denis B et al. Hepatitis rebound after infection with yellow fever virus. Emerging Infectious Diseases. 2019;**25**(6):1248-1249. DOI: 10.3201/eid2506.190069

[29] Casadio L et al. Late-onset relapsing hepatitis associated with yellow fever.

New England Journal of Medicine. 2020;**382**:2059-2061. DOI: 10.1056/NEJMc1913036

[30] Oliveira SRM et al. Histopathological changes and immune response profile in the brains of non-human primates naturally infected with yellow fever virus. Viruses. 2025;**17**(3):386. DOI: 10.3390/v17030386

[31] Fradico JRB et al. Serum soluble mediators as prognostic biomarkers for morbidity, disease outcome, and late-relapsing hepatitis in yellow fever patients. Clinical Immunology. 2023;**251**:109321. DOI: 10.1016/j.clim.2023.109321

[32] Quaresma JA et al. Hepatocyte lesions and cellular immune response in yellow fever infection. Transactions of the Royal Society of Tropical Medicine and Hygiene. 2007;**101**(2):161-168. DOI: 10.1016/j.trstmh.2006.02.019

[33] Fernandez-Garcia MD et al. Vaccine and wild-type strains of yellow fever virus engage distinct entry mechanisms and differentially stimulate antiviral immune responses. MBio. 2016;7(1):e01956-e01915. DOI: 10.1128/mBio.01956-15

[34] Melchjorsen J et al. Expression and function of chemokines during viral infections: From molecular mechanisms to in vivo function. Journal of Leukocyte Biology. 2003;**74**:331-343. DOI: 10.1189/jlb.1102577

[35] ter Meulen J et al. Activation of the cytokine network and unfavorable outcome in patients with yellow fever. The Journal of Infectious Diseases. 2004;**190**(10):1821-1827. DOI: 10.1086/425016

[36] Cong Y et al. Characterization of yellow fever virus infection of human and non-human primate antigen presenting cells and their interaction with CD4+ T cells. PLoS Neglected Tropical Diseases. 2016;**10**(5):e0004709. DOI: 10.1371/journal.pntd.0004709

[37] Douam F, Ploss A. Yellow fever virus: Knowledge gaps impeding the fight against an old foe. Trends in Microbiology. 2018;**26**(11):913-928. DOI: 10.1016/j.tim.2018.05.012

[38] Woodson SE, Holbrook MR. Infection of hepatocytes with 17-D vaccine-strain yellow fever virus induces a strong proinflammatory host response. The Journal of General Virology. 2011;**92**(Pt 10):2262-2271. DOI: 10.1099/vir.0.031617-0

[39] Woodson SE, Freiberg AN, Holbrook MR. Differential cytokine responses from primary human Kupffer cells following infection with wild-type or vaccine strain yellow fever virus. Virology. 2011;**412**(1):188-195. DOI: 10.1016/j.virol.2011.01.012

[40] Li G, Duan T, et al. Yellow fever virus infection in Syrian golden hamsters: Relationship between cytokine expression and pathologic changes. International Journal of Clinical and Experimental Pathology. 2008;**1**(2):169-179

[41] Meier KC et al. A mouse model for studying viscerotropic disease caused by yellow fever virus infection. PLoS Pathogens. 2009;5(10):e1000614. DOI: 10.1371/journal.ppat.1000614

[42] Murai M et al. Active participation of CCR5(+)CD8(+) T lymphocytes in the pathogenesis of liver injury in graft-versus-host disease. The Journal of Clinical Investigation. 1999;**104**(1):49-57. DOI: 10.1172/JCI6642

[43] Hudson NP. The pathology of experimental yellow fever in the Macacus

rhesus. The American Journal of Pathology. 1928;**4**:395-429

[44] Klotz O, Belt TH. The pathology of liver in yellow fever. The American Journal of Pathology. 1930;**6**:663-687

[45] Klotz O, Belt TH. Regeneration of liver and kidney following yellow fever. The American Journal of Pathology. 1930;**6**:689-697

[46] Smetana P. Fifth west Australian random sample egg laying test: 1961-1962. Journal of the Department of Agriculture. 1962;**3**:273-276. Available from: https://library.dpird.wa.gov.au/journal_agriculture4/vol3/iss4/5

[47] Vieira WT et al. Histopathology of the human liver in yellow fever with special emphasis on the diagnostic role of the councilman body. Histopathology. 1983;**7**:195-208

[48] Branquet D. A yellow fever liver. La Medicina Tropical. 1996;**56**:224-227

[49] Quaresma JAS et al. Reconsideration of histopathology and ultrastructural aspects of the human liver in yellow fever. Acta Tropica. 2005;**94**:116-127

[50] Quaresma JA et al. Immunohistochemical examination of the role of Fas ligand and lymphocytes in the pathogenesis of human liver yellow fever. Virus Research. 2006;**116**(1-2):91-97. DOI: 10.1016/j.virusres.2005.08.019

[51] Quaresma JA, Duarte MI, Vasconcelos PF. Midzonal lesions in yellow fever: A specific pattern of liver injury caused by direct virus action and in situ inflammatory response. Medical Hypotheses. 2006;**67**(3):618-621. DOI: 10.1016/j.mehy.2006.01.060

[52] Watson AM, Klimstra WB. T cell-mediated immunity towards yellow fever virus and useful animal models. Viruses. 2017;**9**(4):77. DOI: 10.3390/v9040077

[53] Quaresma JA et al. Revisiting the liver in human yellow fever: Virus-induced apoptosis in hepatocytes associated with TGF-beta, TNF-alpha and NK cells activity. Virology. 2006;**345**(1):22-30. DOI: 10.1016/j.virol.2005.09.058

[54] de Ávila RE et al. Clinical profiles and factors associated with mortality in adults with yellow fever admitted to an intensive care unit in Minas Gerais. The Brazilian Journal of Infectious Diseases. 2020;**93**:90-97. DOI: 10.1016/j.ijid.2020.01.039

[55] Monath TP. Yellow fever: A medically neglected disease. Report on a seminar. Reviews of Infectious Diseases. 1987;**9**(1):165-175. DOI: 10.1093/clinids/9.1.165

[56] PAHO/WHO. Laboratory Diagnosis of Yellow Fever Virus Infection. 2018. Available from: https://www.paho.org/en/documents/laboratory-diagnosis-yellow-fever-virus-infection [Accessed: May 01, 2025]

[57] Vasconcelos PFC, Quaresma JAS. Yellow fever. In: Mehlhorn H, Heukelbach J, editors. Infectious Tropical Diseases and One Health in Latin America. Cham: Springer International Publishing; 2022. Parasitology Research Monographs. pp. 1-17. DOI: 10.1007/978-3-030-99712-0_1

[58] World Health Organization. Surveillance Standards for Vaccine-Preventable Diseases. Yellow Fever [Internet]. Geneva: World Health Organization (WHO); 2020. Available from: https://www.who.int/publications/m/item/vaccine-preventable-diseases-surveillance-standards-yellow-fever [Accessed: May 11, 2025]

[59] Niedrig M et al. Assessment of IgG antibodies against yellow fever virus after vaccination with 17D by different assays: Neutralization test, haemagglutination inhibition test, immunofluorescence assay and ELISA. Tropical Medicine and International Health. 1999;**4**(12):867-871. Available from: https://onlinelibrary.wiley.com/doi/abs/10.1046/j.1365-3156.1999.00496.x

[60] Wieten RW et al. A single 17D yellow fever vaccination provides lifelong immunity; characterization of yellow-fever-specific neutralizing antibody and T-cell responses after vaccination. PLoS One. 2016;**11**(3):e0149871. DOI: 10.1371/journal.pone.0149871

[61] Rasulova M et al. A high-throughput yellow fever neutralization assay. Microbiology Spectrum. 2022;**10**(3):e02548-e02521. DOI: 10.1128/spectrum.02548-21

[62] Brasil. Ministério da Saúde. Secretaria de Vigilância em Saúde. Departamento de Imunização e Doenças Transmissíveis. Manual de manejo clínico da febre amarela [Internet]. 1st ed. Brasília: Ministério da Saúde; 2020. p. 56. ISBN 978-85-334-2818-8. Available from: http://bvsms.saude.gov.br/bvs/publicacoes/manual_manejo_clinico_febre_amarela.pdf [Accessed: May 21, 2025]

[63] Rezende IM et al. Sofosbuvir off-label treatment of yellow fever patients during an outbreak in Brazil, 2018: A cohort study. Open Forum Infectious Diseases. 2024;**11**(6):ofae312. DOI: 10.1093/ofid/ofae312

[64] Possas C et al. Urgent call for action: Avoiding spread and re-urbanisation of yellow fever in Brazil. Memórias do Instituto Oswaldo Cruz. 2018;**113**(1):1-2. DOI: 10.1590/0074-02760170361

[65] Possas C et al. Yellow fever outbreak in Brazil: The puzzle of rapid viral spread and challenges for immunization. Memórias do Instituto Oswaldo Cruz. 2018;**113**(10):e180278. DOI: 10.1590/0074-02760180278

[66] Beasley DWC et al. Yellow fever virus: Genetic and phenotypic diversity and implications for detection, prevention and therapy. Antiviral Research. 2015;**115**:48-70. DOI: 10.1016/j.antiviral.2014.12.010

[67] Monath TP. Yellow fever vaccine. Expert Review of Vaccines. 2005;**4**(4):553-574. DOI: 10.1586/14760584.4.4.553

[68] Galler R et al. The yellow fever 17D vaccine virus: Molecular basis of viral attenuation and its use as an expression vector. Brazilian Journal of Medical and Biological Research. 1997;**30**(2):157-168. DOI: 10.1590/S0100-879X1997000200002

[69] Pulendran B. Learning immunology from the yellow fever vaccine: Innate immunity to systems vaccinology. Nature Reviews Immunology. 2009;**9**(10):741-747. DOI: 10.1038/nri2629

[70] Xeyla. Brazil Resumes Export of Yellow Fever Vaccines. 2019. Available from: https://portal.fiocruz.br/en/news/brazil-resumes-export-yellow-fever-vaccines [Accessed: April 12, 2025]

[71] Chang GJ et al. Nucleotide sequence variation of the envelope protein gene identifies two distinct genotypes of yellow fever virus. Journal of Virology. 1995;**69**(9):5773-5780. DOI: 10.1128/jvi.69.9.5773-5780.1995

[72] Davis EH, Barrett ADT. Structure-function of the yellow fever virus envelope protein: Analysis of antibody epitopes. Viral Immunology.

2020;**33**(1):12-21. DOI: 10.1089/vim.2019.0107

[73] Wieten RW et al. Comparison of the PRNT and an immune fluorescence assay in yellow fever vaccinees receiving immunosuppressive medication. Vaccine. 2016;**34**(10):1247-1251. DOI: 10.1016/j.vaccine.2016.01.037

[74] Campi-Azevedo AC et al. Booster dose after 10 years is recommended following 17DD-YF primary vaccination. Human Vaccines and Immunotherapeutics. 2016;**12**(2):491-502. DOI: 10.1080/21645515.2015.1082693

[75] Haslwanter D et al. Genotype-specific features reduce the susceptibility of south American yellow fever virus strains to vaccine-induced antibodies. Cell Host and Microbe. 2022;**30**(2):248-259.e6. DOI: 10.1016/j.chom.2021.12.009

[76] Barrett AD, Teuwen DE. Yellow fever vaccine - How does it work and why do rare cases of serious adverse events take place? Current Opinion in Immunology. 2009;**21**(3):308-313. DOI: 10.1016/j.coi.2009.05.018

[77] Pagnon A et al. Next generation yellow fever vaccine induces an equivalent immune and transcriptomic profile to the current vaccine: Observations from a phase I randomised clinical trial. eBioMedicine. 2024;**108**:105332. DOI: 10.1016/j.ebiom.2024.105332

[78] Modjarrad MDK et al. Safety and immunogenicity of a next-generation live-attenuated yellow fever vaccine produced in a Vero cell line in the USA: A phase 1 randomised, observer-blind, active-controlled, dose-ranging clinical trial. The Lancet Infectious Diseases. 2024;**24**(12):1393-1402. DOI: 10.1016/S1473-3099(24)00406-7

[79] Montalvo Zurbia-Flores G, Rollier CS, Reyes-Sandoval A. Re-thinking yellow fever vaccines: Fighting old foes with new generation vaccines. Human Vaccines and Immunotherapeutics. 2022;**18**(1):1895644. DOI: 10.1080/21645515.2021.1895644

Chapter 3

# Humoral Immune Response to Dengue Virus Serotypes: Challenges and Emerging Solutions

*Martha Pedraza-Escalona, Desiree Izar-Marín, Juan Carlos Muñoz-Herrera and Sonia Mayra Pérez-Tapia*

## Abstract

Dengue virus (DENV), a mosquito-borne pathogen comprising four distinct serotypes (DENV-1 to DENV-4), poses a significant public health challenge in tropical and subtropical regions, with an estimated 390 million infections occurring annually. The complexity of the immune response to dengue virus is underscored by the dual role of antibodies: while they can provide serotype-specific protection, cross-reactive or sub-neutralizing antibodies may mediate antibody-dependent enhancement (ADE)—a key driver of severe disease outcomes such as dengue hemorrhagic fever (DHF) and dengue shock syndrome (DSS). This review explores the molecular and immunological basis of humoral responses to dengue virus, focusing on antibodies targeting structural proteins and nonstructural protein 1 (NS1). Antibodies directed against EDIII exhibit potent neutralizing activity with minimal ADE risk, whereas antibodies targeting the membrane protein and the EDII fusion loop (FL) are often cross-reactive and can enhance infection *via* Fcγ receptor-mediated pathways. Anti-NS1 antibodies, while capable of providing protection, have also been implicated in pathogenesis through mechanisms involving complement activation and molecular mimicry. The phenomenon of ADE is not exclusive to dengue; cross-reactive antibodies may also worsen infections caused by other flaviviruses, such as Zika virus (ZIKV). To mitigate these risks, antibody engineering strategies—including Fc region modifications and bispecific antibody constructs—are being explored to enhance neutralizing capacity while reducing ADE risk. Vaccine development against dengue has progressed across multiple platforms, including live-attenuated vaccines, inactivated formulations, subunit vaccines, virus-like particles (VLPs), viral vectors, and nucleic acid-based approaches. These strategies aim to enhance humoral immunity and inhibit multiple stages of disease progression.

**Keywords:** antibody-dependent enhancement, dengue virus, humoral response, neutralizing antibodies, vaccine development

## 1. Introduction

Dengue is a mosquito-borne viral disease, classified into four serotypes (DENV-1 to DENV-4) and subdivided into distinct genotypes, that is endemic in tropical and subtropical regions worldwide, where nearly half of the world's population resides. This virus infection results in an estimated 390 million cases annually, with approximately 96 million manifesting clinically, encompassing a spectrum from mild manifestations to potentially fatal cases [1, 2]. Dengue virus (DENV) belongs to the Flaviviridae family and genus Flavivirus, transmitted by mosquitoes of the genus Aedes, primarily *A. aegypti* and, to a minor degree, *A. albopictus* [3]. Along with DENV, this genus also includes a number of other medically important viruses such as West Nile virus (WNV), yellow fever virus (YFV), Japanese encephalitis virus (JEV), and Zika virus (ZIKV) [4]. Dengue is characterized by headache, myalgia, arthralgia, and severe retro-orbital pain. In addition, hemorrhagic manifestations are associated with thrombocytopenia and plasma leakage, which may lead to hypotensive shock and death [5].

The World Health Organization (WHO) documented that from 2000 to 2019, there was a 10-fold surge in reported cases worldwide; the year 2019 marked an unprecedented peak, with reported instances spreading across 129 countries [2]. Since 2023, a global surge in dengue cases has been reported, marked by a significant increase in both the number and magnitude of outbreaks with a predominance of the DENV-3 serotype, as well as their simultaneous occurrence across multiple regions, including areas previously unaffected by dengue [2]. Indeed, the Pan American Health Organization (PAHO) reported more than 13,000 cases with this serotype in the Americas Region in 2024 [6]. These trends are largely driven by rapid urbanization, the expanding geographic range of the vector, accelerated population growth, and the escalating impacts of climate change [7]. The frequent occurrence of outbreaks and the high burden of dengue disease place substantial pressure on healthcare systems and national economies.

The humoral immune response to DENV infection involves a complex interplay among different antibody classes, their temporal dynamics, and their specificity for viral antigens, all of which are crucial for both viral clearance and long-term protection [8]. However, in the case of dengue, the humoral immune response plays a dual role; this duality creates a complex immunological landscape in which neutralizing antibodies confer serotype-specific protection, while non-neutralizing or suboptimal antibodies can mediate sub-neutralizing interactions that facilitate viral entry into Fc receptor-bearing cells in secondary infections, a mechanism known as antibody-dependent enhancement (ADE) [8, 9]. This phenomenon has been strongly associated with increased risk of severe dengue manifestations, such as dengue hemorrhagic fever (DHF) and dengue shock syndrome (DSS) [10, 11]. During primary infection with a single DENV serotype, individuals typically develop lifelong immunity to the homologous serotype through the production of serotype-specific neutralizing antibodies [12, 13]. However, most of the antibody response consists of cross-reactive antibodies that exhibit low or no neutralizing activity, which may contribute to ADE during secondary infection with a heterologous serotype [13–15].

Understanding the specificity, durability, and functional quality of antibody responses is essential to elucidate mechanisms of immune protection versus ADE. This knowledge is imperative for the rational vaccine design, the development of safe and effective therapeutics—including monoclonal antibodies (mAbs) in preclinical and clinical trials [16–19]—and the identification of improved correlates of protection or risk. This work summarizes the current knowledge of the humoral immune response to DENV infection, encompassing its protective functions, its central role in ADE and pathogenesis, key viral antigen targets, recent advances in characterization of antibody repertoires and epitopes, and the implications for vaccine development and clinical management.

## 2. Antibody specificity and epitope recognition

DENV possesses a lipid envelope and a positive-sense single-stranded RNA genome with a single open-reading frame (ORF) that encodes three structural proteins—envelope glycoprotein (E), nucleocapsid protein (C), and membrane protein (prM/M)—as well as seven nonstructural proteins (NS1, NS2A, NS2B, NS3, NS4A, NS4B, and NS5) (**Figure 1**) [20]. The C protein encapsulates the viral genome,

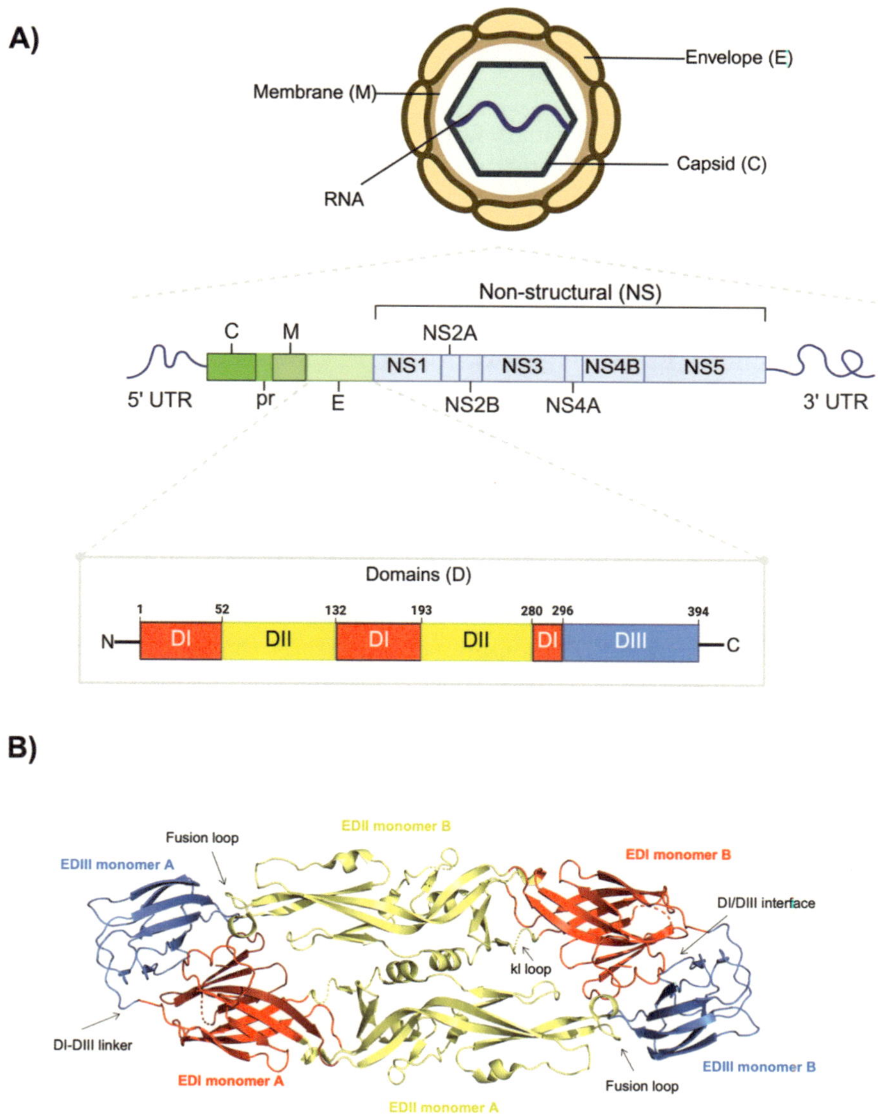


**Figure 1.**
*General characteristics of dengue virus. (A) Dengue virus possesses a positive-sense RNA genome (approximately 10.6 kb) with a single open-reading frame (ORF), flanked by 5′ and 3′ untranslated regions (UTRs). This ORF encodes a polyprotein that is cleaved by both viral and host proteases into three structural proteins: envelope glycoprotein (E), nucleocapsid protein (C), and membrane protein (prM/M), and seven nonstructural proteins: NS1, NS2A, NS2B, NS3, NS4A, NS4B, and NS5 (sky blue). Created in BioRender. https://BioRender.com/l5esqws. (B) The E protein is the major target of the immune response and consists of three domains: EDI (red), EDII (yellow), and EDIII (marine blue). In the mature virion, the E protein is organized as 90 antiparallel dimers on the viral surface. E protein PDB: 5NoA (visualized by PyMOL v 2.5.4).*

which is enclosed within a lipid bilayer containing the E and M proteins [21]. The E and prM proteins form the viral glycoprotein embedded in the lipid bilayer [22, 23]. The prM protein facilitates proper folding and assembly of the E protein and is presented on the surface of partially mature virions [24]. Among these proteins, E and the prM—both exposed on the virion surface—are the primary targets of the host immune response during DENV infection [22, 25]. Additionally, NS1, which is involved in viral replication and packaging [24], is also recognized by the immune system, although with lower intensity and frequency [26].

### 2.1 Antibodies targeting E protein

The DENV E protein is the primary target of neutralizing antibody responses following dengue infection. In the mature virion, the E protein is organized as 90 antiparallel dimers on the viral surface and mediates attachment to host cell receptors, membrane fusion, and viral entry *via* receptor-mediated endocytosis (**Figure 1**) [27]. Each E protein monomer comprises three domains: a central β-barrel domain (EDI), an elongated dimerization domain (EDII) containing the fusion loop (FL) at its distal end—critical for initiating fusion—and a carboxy-terminal immunoglobulin-like domain (EDIII) (**Figure 1**) [28, 29]. Due to approximately 62–80% residue sequence identity among the E proteins of the four serotypes (**Figure 2**) [30], antibody response can be either serotype-specific or cross-reactive. The dominant antibody repertoire in human serum targets the fusion loop in EDII (DII-FL epitope). These antibodies are broadly cross-reactive across the four DENV serotypes and ZIKV, but they exhibit limited neutralizing activity and can induce ADE [12, 13, 31]. Although EDI is not a major target of neutralizing antibodies, it can elicit humoral responses, particularly during secondary dengue infections. Antibodies targeting EDI are generally broad, cross-reactive, and have low neutralizing capacity [32]. In contrast, antibodies against EDIII comprise only 5–10% of the total humoral response in infected individuals [33], but they tend to be strongly neutralizing against specific serotypes, though not uniformly across all four [29]. Additionally, a small subset of antibodies has been identified that recognize different conformational epitopes spanning multiple domains in EDI, EDII, and EDIII, known as the E-dimer epitope (EDE), which can be classified into two types: (1) those whose binding depends on the presence of an N-linked glycan at position N154 of the E protein and (2) those that bind in the absence of this glycan [34, 35]. These antibodies can be broadly neutralizing, with some exhibiting serotype-specificity and others cross-reactive [36].

### 2.2 Antibodies targeting prM

The prM protein is composed of 166 residues (approximately 19 kDa). During viral morphogenesis, it is cleaved by the host protease furin, releasing a 91-amino acid N-terminal fragment known as the pr-peptide. This cleavage generates the mature M protein, comprising residues 92 to 166. The pr-peptide acts as a protective lid over the hydrophobic fusion loop of the E protein, shielding the fusogenic region and thereby preventing premature membrane fusion during the transport of immature virions through the host secretory pathway [37].

The prM protein is a potent inducer of the immune response. Analysis of the memory B-cell responses in DENV-infected individuals has shown that approximately 60% of the human antibody response targets the prM protein [32], including a subset

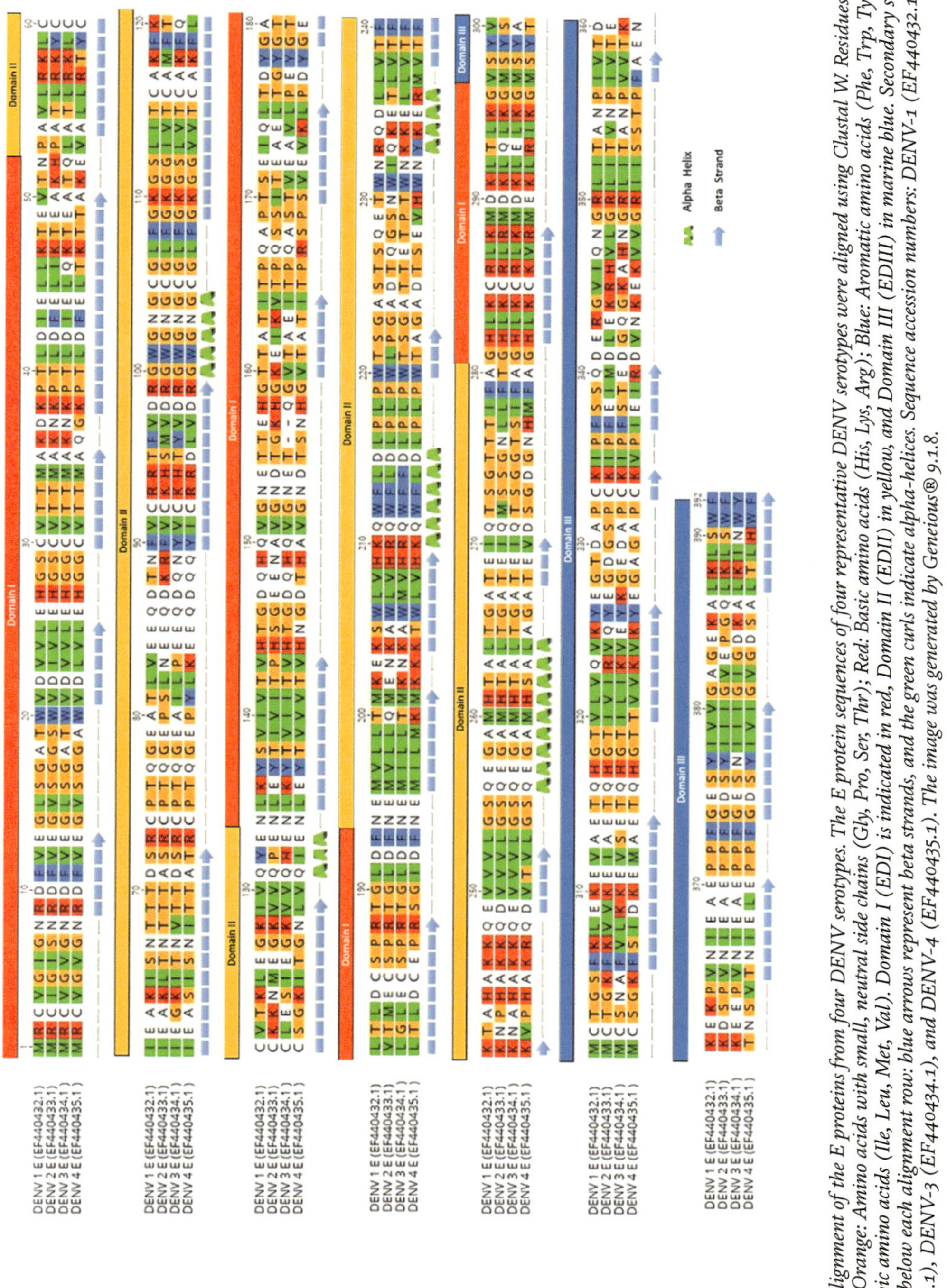


**Figure 2.**
*Sequence alignment of the E proteins from four DENV serotypes. The E protein sequences of four representative DENV serotypes were aligned using Clustal W. Residues are color-coded as follows: Orange: Amino acids with small, neutral side chains (Gly, Pro, Ser, Thr); Red: Basic amino acids (His, Lys, Arg); Blue: Aromatic amino acids (Phe, Trp, Tyr); and Green: Hydrophobic amino acids (Ile, Leu, Met, Val). Domain I (EDI) is indicated in red, Domain II (EDII) in yellow, and Domain III (EDIII) in marine blue. Secondary structure elements are shown below each alignment row: blue arrows represent beta strands, and the green curls indicate alpha-helices. Sequence accession numbers: DENV-1 (EF440432.1), DENV-2 (EF440433.1), DENV-3 (EF440434.1), and DENV-4 (EF440435.1). The image was generated by Geneious® 9.1.8.*

with neutralizing activity [38]. However, several studies have demonstrated that prM elicits highly cross-reactive antibodies capable of recognizing prM across all four DENV serotypes. Notably, anti-prM antibodies generally exhibit poor neutralizing capacity but can strongly enhance infection through ADE [10, 39, 40]. Interestingly, D29 Fab-IgG, a DENV cross-reactive monoclonal antibody targeting prM, has been shown to bind with high affinity to a conformational epitope and to restore the infectivity of otherwise noninfectious immature DENV particles in K562 cells [41].

### 2.3 Antibodies targeting NS1

NS1 is a highly immunogenic nonstructural protein involved in both immune protection and dengue pathogenesis. It is a glycoprotein with a molecular weight ranging from 46 to 55 kDa, depending on its glycosylation status. NS1 can be anchored to the surface of the infected cells, associated with intracellular vesicular compartments, or secreted into the extracellular space as oligomers [42]. Antibodies targeting NS1 have been shown to confer protection against dengue infection [43]. These antibodies can activate the complement system, thereby contributing to immunity defense by promoting the clearance of infected cells and interfering with NS1-mediated immune evasion mechanisms [44]. However, certain anti-NS1 antibodies exhibit molecular mimicry and cross-react with host proteins such as ATP synthase, vimentin, or monocyte chemotactic protein-1. This cross-reactivity may contribute to endothelial dysfunction, platelet depletion, and vascular leakage—key features of severe dengue disease [45–47]. Despite the dual role of NS1 in the immune response, it remains a critical diagnostic marker, as it can be detected in the bloodstream during the early stages of dengue infection [48].

### 2.4 Humoral memory response in dengue infection

Several reports have shown that IgM antibodies are produced early during primary infection and are typically short-lived. In contrast, IgG antibodies (predominantly IgG1 and IgG3) emerge later [49] and persist for extended periods. Memory B cells specific to DENV have been reported to remain stable for several years postinfection, comprising both serotype-specific and cross-reactive clones. Remarkably, some reports suggest that immunological memory against DENV can persist for up to 60 years [50]. Upon secondary DENV exposure, memory B cells differentiate into antibody-secreting cells, leading to the switch production of IgG antibodies, often preceding detectable viremia [51]. The quality of this memory response, including its breadth and neutralizing capacity, is a key determinant of disease outcome.

## 3. Antibody-dependent enhancement

Infection with any of the four DENV serotypes can lead to a wide spectrum of clinical symptoms, ranging from asymptomatic or mild dengue fever (DF) to severe and potentially fatal forms, including DHF and DSS. Fortunately, severe manifestations remain relatively rare. A primary infection with one serotype elicits a strong and long-lasting serotype-specific (homotypic response) antibody response but confers only transient cross-protection against the other serotypes (heterotypic response) [52]. Despite the breadth of antigenic recognition, these antibodies typically prevent infection only by the homologous serotype [53, 54]. Conversely, secondary infection

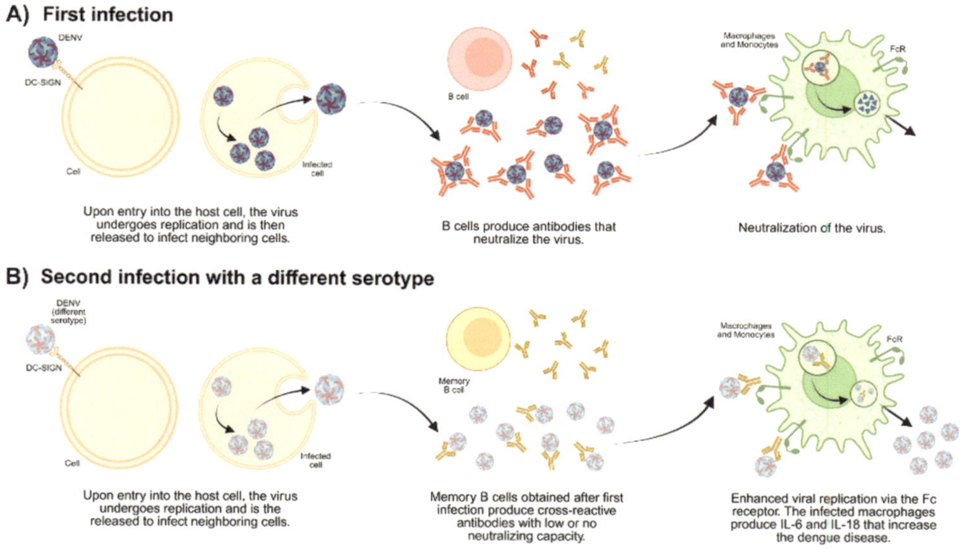


**Figure 3.**
*Antibody-dependent enhancement in DENV infection. (A) During primary infection, the humoral immune response produces both neutralizing antibodies and non-neutralizing antibodies from different B cells. Neutralizing antibodies can inhibit the virus replication. (B) During secondary infection with a different DENV serotype, pre-existing non-neutralizing antibodies, generated by memory B cells after the first infection, form complexes with the virus and facilitate its entry into susceptible macrophages and monocytes via Fcγ receptors, allowing viral replication. Infected cells subsequently produce cytokines such as IL-6, IL-8, and IL-10, which contribute to increased disease severity. Created in BioRender. https://BioRender.com/ql2p229.*

with a different serotype not only provides limited protection but also increases risk factors for developing DHF or DSS [55, 56], primarily due to a phenomenon known as antibody-dependent enhancement (ADE) (**Figure 3**) [53, 57].

According to the ADE hypothesis, during a secondary infection with a heterologous DENV serotype, the preexisting cross-reactive and weakly neutralizing antibodies from the primary infection bind to the new serotype without preventing or clearing the infection. These antibody-virus complexes are then recognized by the Fc gamma receptor (FcγR) of phagocytic cells such as monocytes, macrophages, and dendritic cells, facilitating viral entry and replication and leading to increased viremia [9, 57]. Specifically, the Fc region of the antibody interacts with the FcγR, while the Fab portion binds the viral antigen, enabling internalization of the virus-antibody complex [56]. When the affinity between IgG and FcγR is low, the immune complex dissociates within the cell, shifting the immune response from antiviral to immunosuppressive. In fact, the ADE activity has been ranked in accordance with the subclass of IgG: IgG3 > IgG1 > IgG2 > IgG4 [58, 59]. This process suppresses the production of key antiviral mediators such as IL-12, IFN-γ, and nitric oxide, while upregulating IL-6, IL-8, and IL-10. The imbalance in cytokine signaling facilitates uncontrolled viral DENV replication, vascular leakage, and endothelial dysfunction—hallmarks of severe dengue disease [60–63]. The severity of these outcomes is influenced not only by viral load but also by the specificity and concentration of the antibodies involved.

Most antibodies that target EDIII generate a neutralizing response, exhibiting low ADE potential [59]. In contrast, antibodies directed against prM and the EDII FL region are highly cross-reactive across DENV serotypes but poorly neutralizing, making them potent ADE promoters [39]. Interestingly, anti-NS1 antibodies also contribute to disease severity through ADE-like mechanisms, particularly *via* complement activation and endothelial damage [61]. These antibodies can cross-react with host proteins, including endothelial components and platelets, promoting vascular leakage and thrombocytopenia through molecular mimicry [64]. Moreover, NS1 and anti-NS1 immune complexes can activate complement pathways, amplifying inflammation and contributing to tissue damage [65].

Due to the high degree of sequence identity among flaviviruses, cross-reactive immune responses also play a role in ADE during secondary infections involving related viruses. For example, DENV-specific monoclonal antibodies (mAbs) have been shown to enhance ZIKV infection [66]. mAbs targeting the fusion loop epitope were able to bind ZIKV but failed to neutralize it, instead promoting ADE [67]. Further, *in vivo* studies in mice demonstrated that some antibodies targeting domains I/II of the ZIKV E protein can cross-react with DENV, exacerbating disease severity [31]. Bardina et al. [68] showed that convalescent plasma from DENV-infected individuals enhanced ZIKV infection in susceptible mice, increasing fever, viremia, morbidity, and mortality. *In vitro*, DENV and WNV antisera also enhanced ZIKV infection in K562 cells, highlighting the potential of cross-reactive flavivirus antibodies to drive ADE across species boundaries [69].

## 4. Advances and applications of antibody engineering against DENV

One of the main challenges in dengue research is the development of effective therapeutic and prophylactic strategies that circumvent the risks associated with ADE. In this context, antibody engineering has emerged as a powerful approach to optimize specificity, enhance neutralization potency, and improve safety profiles. Several engineered antibody formats—such as single-chain fragment variable (scFv), antigen-binding fragments (Fab), Fc-modified antibodies, and bispecific constructs—are currently under investigation. Among these strategies, Fc region modifications offer a promising means to modulate or eliminate FcγR-Fc interactions. For example, a deletion comprising residues 231–239 at the N terminus of the CH2 domain—within the "lower hinge" region involved in the binding of FcγRI, FcγRII, and C1q—has been shown to abrogate enhancing activity [70]. Similarly, substituting asparagine with glutamine or alanine at position 297 in the Fc of human mAbs has conferred broad neutralizing activity against all four DENV serotypes without inducing ADE [71]. Another well-characterized modification involves double alanine substitutions at positions 234 and 235 (LALA mutation), which effectively disrupt FcγR binding while preserving antigen recognition [30]. Williams et al. [72] showed that humanized mAbs incorporating LALA mutations retained potent neutralizing capacity and abolished enhancement of infection *in vitro* and *in vivo*. Furthermore, LALA variants of broadly neutralizing antibodies targeting the DENV E-dimer epitope have shown protective efficacy against ZIKV in mouse and nonhuman primate models, without triggering ADE (**Figure 4**) [34].

Bispecific antibodies (bsAbs), which are engineered to recognize two different epitopes on the same or different antigens, offer another innovative strategy. In the context of DENV, bsAbs can simultaneously target two neutralizing epitopes across different serotypes, thereby enhancing breadth and potency. For instance, Shi et al. [73] developed a bispecific antibody (DVD-1A1D-2A10) capable of binding both E-DIII and E-DII-FL of the DENV E protein. This dual-targeting strategy significantly improved neutralization across all four DENV serotypes. To eliminate FcγR interaction and prevent ADE, the Fc region of the antibody was engineered by

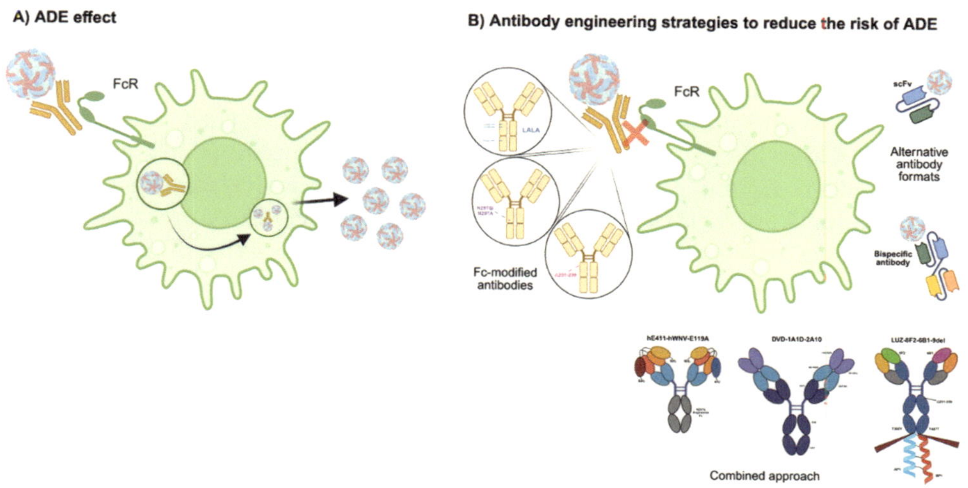


**Figure 4.**
*Antibody engineering strategies for dengue virus control. (A) Antibody-dependent enhancement (ADE) is mediated by pre-existing non-neutralizing IgG antibodies that form immune complexes with the DENV, facilitating its entry into the host cell and promoting viral replication. (B) Several antibody engineering strategies have been developed to improve specificity, enhance neutralizing potency, and reduce the risk of ADE:* Strategy 1: Fc-modified antibodies. *These modifications aim to prevent binding to the Fcγ receptors (FcγRI, FcγRII, and C1q): (a) LALA mutations at positions 234 and 235 located in the CH2 domain and P329G in the CH3 domain; (b) N297Q/N297A mutations in the CH2 domain; and (c) deletion of residues 231 to 239 at the N-terminal of the CH2 domain (lower hinge region).* Strategy 2: Alternative antibody formats. *The use of single-chain fragment variables (scFvs) and bispecific antibodies (bsAbs) to improve targeting and functionality.* Strategy 3: Combined approach. *Integration of Strategies 1 and 2 through the development of Fc-modified bsAbs, including DVD-1A1D-2A10 [73], h4E11–hWNV-E119A N297Q [74], and LUZ-8F2-6B1-9del [35]. See more information in the text. Created in BioRender. https://BioRender.com/a5ll4hf.*

deleting nine residues at the N terminus (positions 231–239). Additional examples include two bispecific tetravalent antibodies incorporating the N297Q mutation: an Ig-DART (4E11-E60) [75] and an Fc-DART (h4E11–hWNV-E119A N297Q). The latter demonstrated protective activity against DENV-2 and DENV-3 in a mouse model [74]. LUZ-8F2-6B1-9del is a bispecific antibody generated by combining the 6B1 mAb, which recognizes DENV-1, 2, and 3, with the 8F2 mAb, which targets DENV-4. The construct incorporates two leucine zipper peptides, Ap1 and Bp1, to promote heterodimerization of half antibodies and includes a deletion of residues 231–239 in the Fc region to prevent FcγR binding. This bsAb was capable of neutralizing all four DENV serotypes and effectively abrogating ADE (**Figure 4**) [35].

Despite these promising results, the broader immunological consequences of Fc modifications remain unclear, particularly their potential effects on other Fc-mediated immune functions such as complement activation and antibody-dependent cell-mediated cytotoxicity (ADCC). Further studies are needed to fully evaluate the safety and functional implications of these engineered antibodies.

## 5. Vaccines designed against DENV

Vaccination remains the most effective strategy for preventing dengue infection. However, the development of a dengue vaccine is particularly challenging due to the need to induce balanced and long-lasting immunity against all four DENV serotypes. Despite decades of intense research, progress has been hindered by several factors, including the risk of ADE, serotype variability, and complex immunopathogenesis mechanisms. The primary objectives of dengue vaccine development are twofold: to elicit robust, serotype-balanced neutralizing antibody responses and to avoid immunopathological outcomes associated with suboptimal immunity. To date, only two live-attenuated vaccines—Dengvaxia® and Qdenga®—have been approved for use in specific populations within dengue-endemic regions. A variety of other vaccine platforms—including inactivated, subunit, virus-like particle (VLP), viral vector, and nucleic acid-based approaches—are currently under investigation in both preclinical and clinical stages (see **Table 1**).

### 5.1 Live-attenuated vaccines

Live-attenuated vaccines are developed by reducing the virulence of the pathogen while preserving its ability to replicate. These vaccines typically induce strong and long-lasting immunity and are relatively cost-effective to produce.

Dengvaxia® (CYD-TDV), developed by Sanofi Pasteur, was the first licensed dengue vaccine. It is a tetravalent live-attenuated chimeric vaccine based on the yellow fever 17D backbone, into which the prM and E genes of each DENV serotype have been inserted [76]. Clinical trials demonstrated moderate overall efficacy (~60%), with the highest protection observed against DENV-3 and DENV-4 [77]. However, post-licensure surveillance revealed an increased risk of severe dengue in seronegative individuals following natural infection, likely due to vaccine-induced priming that mimics a primary infection and predisposes to ADE [78]. This was particularly evident in a group of children under 9 years old, who experienced an elevated risk of severe disease upon infection after vaccination [79]. As a result, the WHO recommends Dengvaxia® only for seropositive individuals aged 9–45 years.

Qdenga® (TAK-003), developed by Takeda Vaccines, is the second approved tetravalent live-attenuated dengue vaccine. It is based on a DENV-2 backbone engineered to express the prM-E genes of DENV-1, DENV-3, and DENV-4, creating recombinant chimeric viruses DENV2/1, DENV2/3, and DENV2/4. Unlike Dengvaxia®, this platform does not use yellow fever virus components, potentially reducing the risk of cross-reactive immunopathology. The vaccine elicits a robust humoral response, including the production of neutralizing antibodies against DENV, cellular immune responses, and anti-NS1 antibodies, the latter due to its inclusion of the DENV-2 nonstructural genome [80]. Phase III clinical trials showed an overall efficacy of 80.2%, with protection observed even in seronegative individuals [81]. Additionally, TAK-003 has demonstrated a favorable safety profile across different serostatus groups, suggesting broader applicability. Nevertheless, long-term efficacy data are still being gathered to assess the durability and consistency of protection.

TV003/TV005, developed by the U.S. National Institute of Allergy and Infectious Diseases (NIAID), are three full-length wild-type viruses and one chimeric virus; the latter is based on the DENV-4 backbone with the prM and E genes of DENV-2. These vaccines are derived from attenuated strains by deletion in the 3′ untranslated region (UTR) [82]. Clinical trials have shown promising immunogenicity, including balanced seroconversion against all four serotypes after a single dose [83]. TV003 is undergoing phase III evaluation in endemic regions.

Live-attenuated vaccines have shown considerable promise in controlling dengue fever, particularly when administered in targeted populations and endemic settings, where they have contributed to reductions in disease incidence and severity. However, differences in protective efficacy across DENV serotypes, as well as variability in vaccine effectiveness based on age and preexisting immunity, remain

DOI: http://dx.doi.org/10.5772/intechopen.1012333

| Vaccine platform | Examples | Strategy | Advantages | Main limitations | Challenges |
|---|---|---|---|---|---|
| Live attenuated | Dengvaxia® | YF17D backbone + prM/E of DENV-1 to DENV-4 | Strong and long-lasting immunity; cost-effective; efficient in endemic populations; single-dose seroconversion to all serotypes (TV003/TV005) | Risk of ADE; serotype imbalance; age- and serostatus-dependent efficacy; limited long-term protection data | Improve safety in seronegative individuals by reducing ADE risk (Dengvaxia®); optimize serotype balance and durability of protection; strengthen trial data in endemic regions |
| | Qdenga® | DENV-2 backbone + prM/E of DENV-1, −3, and − 4 | | | |
| | TV003/TV005 | Attenuated strains DENV-1, −3, −4; chimera (DENV-4 backbone + DENV-2 prM/E) | | | |
| Inactivated virus | DPIV, TPIV | Tetravalent formalin- or AMT-inactivated virus | Enhanced safety; induce balanced antibody response across all serotypes; rapid development | Weak immunogenicity; requires multiple booster doses and adjuvants to generate long-term immunity; potential ADE risk if neutralizing antibody titers decline; no immunity to nonstructural proteins | Improve adjuvant formulations; combine with other platforms to enhance immunogenicity |
| Subunit | V180 | Recombinant proteins based on prM and 80% of E protein (E80) of DENV-1 to −4 combined with various adjuvants | Safe; easy to produce; scalable production; induces balanced immunity across all serotypes with reduced ADE risk | High cost; complex protein folding and stabilization needed; needs multiple doses and requires strong adjuvants for immunogenicity | Optimize antigen design and expression systems to lower costs; enhance adjuvant synergy; and enhance humoral immunity durability |
| Virus-like particles (VLPs) | DSV4 | VLPs expressing EDIII domains of DENV-1 to −4 | Safe for immunocompromised individuals; preserves conformational epitopes; induces potent neutralizing antibodies; elicits a balanced immune response | Complex and expensive manufacturing; requires adjuvants | Scale-up production; explore plant/yeast-based systems; develop stabilization technologies (e.g., nanoparticle formulations) |
| Viral vector | rAdV5 | Adenovirus 5 vector expressing EDIII domains of DENV-1 to −4 | Induces both cellular and humoral immunity; low-dose efficacy; potential for single-dose protection | Preexisting anti-vector immunity may reduce efficacy; possible mild side effects | Use fewer common vectors; integrate into heterologous prime-boost regimens |

| Vaccine platform | Examples | Strategy | Advantages | Main limitations | Challenges |
|---|---|---|---|---|---|
| DNA | D1ME100 | Monovalent DNA vaccine encoding DENV-1 prM/E, co-administered with VAXFECTIN adjuvant | Thermostable; safe; low-cost; scalable production; allows rapid serotype customization | Low immunogenicity in humans; requires multiple doses and strong adjuvants; inefficient transfection; specialized delivery equipment needed | Improve delivery methods (e.g., electroporation); optimize codons; develop better adjuvants |
| mRNA | prM-NS1 mRNA | LNP-encapsulated mRNA encoding DENV-2 prM-E-NS1 | Rapid development; strong humoral and cellular responses; flexible design; high neutralizing antibody titers without ADE (E80) | High cost; cold-chain requirement; possible ADE due to prM-NS1 cross-reactivity; limited long-term protection data | Refine antigen design to minimize ADE; explore non-replicating and NS1-focused mRNA constructs |
| | E80-mRNA | LNP-encapsulated mRNA encoding truncated E(E80) + NS1 DENV-2 | | | |

**Table 1.**
*An overview of the main features, development strategies, advantages, limitations, and challenges of DENV vaccine platforms.*

important challenges. Ongoing efforts aim to optimize both efficacy and safety of live-attenuated vaccines to achieve comprehensive protection against all four dengue virus serotypes.

### 5.2 Inactivated vaccines

Inactivated vaccines are produced by chemically or physically inactivating the virus, thereby eliminating the risk of reversion to virulence. These vaccines are known for their high safety profile, rapid development, and ability to induce immune responses against multiple antigens simultaneously. However, they do not express NS proteins and exhibit weaker immunogenicity, a limited capacity to elicit cellular immune responses, and a potential risk of ADE [84]. In the development of inactivated vaccines, key factors include the virus inactivation methods, choice of adjuvants, and potential combination with other vaccine platforms. Four dengue-inactivated vaccines are currently in clinical trial phases: three are tetravalent, purified, and formalin-inactivated vaccines, and one is inactivated using aminoethyl benzene trioxane (AMT) [85]. All have demonstrated acceptable safety and immunogenicity, although multiple doses and adjuvants (e.g., alum) are required to enhance immune responses [86–89]. Due to its lack of replicative capacity, the purified inactivated vaccine (PIV) platform is considered safer for immunocompromised or pediatric populations.

### 5.3 Subunit and virus-like particle vaccines

Subunit vaccines use viral proteins or peptides—such as membrane-associated or structural proteins—to stimulate immune responses in the absence of live pathogens. This approach offers several advantages, including minimal side effects, a favorable safety profile, a balanced immune response against all four dengue virus serotypes, and ease of large-scale production. However, they generally exhibit weak immunogenicity, requiring the use of adjuvants and often requiring complex immunization schedules. Additional challenges include ensuring proper protein folding, achieving accurate post-translational modifications, and stabilizing proteins or peptides to maintain structural integrity and to enhance immunogenicity [90]. Various expression systems have been employed to produce dengue antigens—particularly the E protein, its DIII, and NS1—for subunit vaccine development. These include insect cells (Sf9 and S2), mammalian cells (HeLa, HEK293, Vero, and BHK), yeast (*S. cerevisiae and P. pastoris*), *E. coli,* and transgenic plants (*Nicotiana tabacum and N. benthamiana*) [91]. Among the most advanced candidates, Merck's V180 is a recombinant subunit vaccine, which has entered Phase I clinical trials. V180 consists of truncated E proteins containing 80% of the N-terminal sequence from DENV1–4, produced in S2 insect cells and formulated with aluminum hydroxide as an adjuvant. In rhesus monkeys, the vaccine induced high levels of neutralizing antibodies against all four DENV serotypes and demonstrated protection against viremia following challenge with DENV1–4 [92]. Several other subunit vaccine candidates expressed in different systems remain under investigation.

VLPs are engineered to mimic the structural features of the native virus without containing viral genetic material. They hold strong promise for dengue vaccine development due to their capacity to present conformational epitopes and stimulate robust B-cell responses. The same expression systems used for subunit vaccines can be employed for VLP production. However, large-scale manufacturing and associated costs remain challenging, as only a limited number of viral proteins are capable of self-assembling VLPs. A recently developed VLP-based vaccine candidate, known as

DSV4 (Dengue Subunit Vaccine Tetravalent), shows considerable promise. This candidate comprises the envelope EDIII regions of all four DENV serotypes expressed in *P. pastoris* and linked *via* flexible peptide linkers within a single open-reading frame. In preclinical studies, DSV4 elicited serotype-specific neutralizing antibodies against all four DENV serotypes in mice when it was administered with aluminum hydroxide as an adjuvant [93].

### 5.4 Viral vector vaccines

Viral vector vaccines utilize modified viruses as delivery systems to introduce genetic material from a pathogen into host cells, where it is expressed to stimulate an immune response that closely mimics natural infection. Several viral vectors have been used in dengue vaccine development, including adenovirus (AdV), adeno-associated virus (AAV), lentivirus, measles virus (MV), modified vaccinia Ankara (MVA), alphavirus, and vesicular stomatitis virus (VSV) [94].

A key advantage of viral vector vaccines is their ability to generate strong immune responses at relatively low doses compared to traditional vaccine platforms. For example, a recombinant adenovirus type 5 vaccine (rAdV5) expressing EDIII of DENV1–4 elicited both cell-mediated immune responses and virus-specific neutralizing antibodies in mice [95]. Viral vector platforms have shown significant efficacy, inducing strong humoral and cellular immune responses.

However, limitations remain, including the presence of preexisting immunity to the viral vector, which may diminish vaccine efficacy [96]. Additionally, mild side effects may occur, like those reported for viral vector-based COVID-19 vaccines [97].

### 5.5 DNA and mRNA vaccines

Nucleic acid vaccines, classified as third-generation vaccine technology, include both DNA and mRNA platforms. These vaccines present antigens in their native conformation and can induce robust humoral and cellular immune responses.

DNA vaccines use genetically engineered plasmid DNA to encode antigenic components of the virus (e.g., prM/E or NS1). Once delivered into host cells, the encoded antigens are synthesized and presented on the cell surface to immune responses [98]. DNA vaccines are thermostable—often eliminating the need for refrigeration—facilitating their easier transportation and distribution. They are also relatively easy to manufacture and have demonstrated favorable safety profiles. Moreover, DNA vaccines can be rapidly developed in response to emerging RNA viruses.

Despite these advantages. DNA vaccines are limited by relatively low has limitations, primarily the relatively low immunogenicity in humans, potentially due to inefficient transfection efficiency [99]. For example, the D1ME100 vaccine, which expressed the DENV-1 prM and E proteins, progressed to clinical trials; nevertheless, its immunogenicity in humans has generally been lower than that of live-attenuated vaccines [100]. Different strategies are being explored to enhance efficacy, including optimized delivery methods such as electroporation, codon optimization, and the use of adjuvants.

The success of mRNA-based COVID-19 vaccines [101] has renewed interest in mRNA vaccine platforms for other viral diseases, including dengue. mRNA vaccines encoding DENV structural proteins can be rapidly designed and relatively easily manufactured and can induce strong humoral and cellular immune responses. A recent study demonstrated that lipid nanoparticle (LNP)-formulated mRNA vaccines encoding DENV prM/NS1 elicited potent neutralizing antibody responses and conferred protection in mice [102]. Nonetheless, these vaccines also induced serotype cross-reactive antibodies associated with ADE. In contrast, a modified non-replicable mRNA vaccine encoding EDIII and NS1 encapsulated in LNP elicited neutralizing antibodies that effectively blocked all four DENV serotypes *in vitro* without inducing significant ADE [103]. Despite these advances, further efforts are needed to improve rational antigen design to enhance cross-protection and eliminate the risk of ADE.

### 5.6 Future directions and considerations for dengue vaccines

Currently approved vaccines, such as Dengvaxia® and Qdenga®, face limitations in achieving balanced and durable protection against all four DENV serotypes, particularly in seronegative individuals. To overcome these challenges, dengue vaccination strategies must shift toward personalized and regionally adapted approaches. Considering the geographic variability in serotype prevalence and population immunity, region-specific vaccine formulations or tailored delivery schedules may enhance overall efficacy.

Moreover, combination strategies—such as heterologous prime-boost regimens (e.g., live-attenuated priming followed by mRNA boosting)—are being actively explored to improve immunogenicity and broaden protection. Continued surveillance for breakthrough infections and long-term safety studies, particularly concerning ADE and potential autoimmune events, remains essential.

Developing a safe and effective dengue vaccine is inherently complex due to the immunological intricacies of the virus. Despite the challenges, multiple vaccine candidates are progressing through clinical development, each offering unique advantages. The next generation of dengue vaccines will likely integrate novel technologies—such as mRNA platforms and NS1-based antigens—alongside improved antigen design coupled with advanced delivery systems. A multidisciplinary and integrative approach encompassing virology, immunology, and epidemiology will be essential to effectively reducing the global burden of dengue.

## 6. Conclusions

Dengue virus infection poses a multifaceted challenge to global public health due to its complex virology, immunopathogenesis, and the paradoxical nature of the immune response. While the humoral immune response is essential for viral neutralization and long-term protection, it also contributes to disease exacerbation through ADE. Antibodies targeting structural proteins such as E and prM, as well as the nonstructural NS1, exhibit diverse functional profiles—ranging from potent neutralization to enhancement of infection. The fine specificity, cross-reactivity, and concentration of these antibodies critically influence disease outcomes, particularly during secondary heterotypic infections.

The persistent risk of ADE remains a significant barrier to effective vaccine design. Although live-attenuated vaccines like Dengvaxia® and Qdenga® have demonstrated partial success, their efficacy is variable across DENV serotypes and is influenced by host factors such as prior exposure and age. This underscores the urgent need for safer, broadly protective, and serotype-balanced vaccines. Advances in antibody engineering—including Fc modifications and bispecific constructs—offer promising therapeutic strategies by enhancing viral neutralization while minimizing ADE.

Emerging vaccine technologies, such as mRNA platforms, VLPs, and viral vectors, provide opportunities to overcome the limitations of traditional vaccine platforms (**Table 1**). Nonetheless, the path to a universally safe and effective dengue vaccine will depend on a more comprehensive understanding of immune correlates of protection, mechanisms of viral immune evasion, and determinants of disease severity.

Ultimately, an integrated approach that combines immunology, virology, molecular biology, and epidemiology is critical to guide the design of next-generation vaccines and therapeutics. Ongoing surveillance, personalized vaccination strategies, and in-depth research into the mechanisms of ADE will be critical for reducing the global burden of dengue and preventing future outbreaks of dengue and other flavivirus infections.

## Acknowledgements

We thank Dr. David Jáuregui-Zúñiga for his contribution in drafting and revising the final version of the manuscript. This study was supported by the SECIHTI Grant CBF2023-2024-2209 to MP-E and the grants SIP-IPN 20232368 and 20240761 to SMP-T.

## Conflict of interest

The authors declare no conflict of interest.

## Author details

Martha Pedraza-Escalona[1*], Desiree Izar-Marín[2], Juan Carlos Muñoz-Herrera[2] and Sonia Mayra Pérez-Tapia[2,3,4]

1 SECIHTI-Research and Development in Biotherapeutics Unit, National School of Biological Sciences, National Polytechnic Institute, Mexico City, Mexico

2 Research and Development in Biotherapeutics Unit, National School of Biological Sciences, National Polytechnic Institute, Mexico City, Mexico

3 National Laboratory for Specialized Services of Investigation, Development and Innovation (I+D+i) for Pharma Chemicals and Biotechnological Products, LANSEIDI-FarBiotec-CONACyT, Mexico City

4 Immunology Department, National School of Biological Sciences, National Polytechnic Institute, Mexico City, Mexico

*Address all correspondence to: maria.pedraza@secihti.mx

## References

[1] Bhatt S, Gething P, Brady O, et al. The global distribution and burden of dengue. Nature. 2013;**496**:504-507. DOI: 10.1038/nature12060

[2] WHO. 2025. Available from: https://www.who.int/emergencies/disease-outbreak-news/item/2023-DON498

[3] Lambrechts L, Scott TW, Gubler DJ. Consequences of the expanding global distribution of Aedes albopictus for dengue virus transmission. PLoS Neglected Tropical Diseases. 2010;**4**:e646. DOI: 10.1371/journal.pntd.0000646

[4] Pierson TC, Diamond MS. Flaviviruses. In: Knipe DM, Howley PM, editors. Fields Virology. 6th ed. Philadelphia, PA: Wolters Kluwerand Lippincott Williams & Wilkins; 2013. pp. 747-794

[5] Mestre Tejo A, Toshie Hamasaki D, Mattos Menezes L, Ho Y-L. Severe dengue in the intensive care unit. Journal of Intensive Medicine. 2024;**4**:16-33. DOI: 10.1016/j.jointm.2023.07.007

[6] PAHO. 2025. Available from: https://www.paho.org/sites/default/files/2025-02/2025-feb-7-phe-epi-alert-dengue-final.pdf

[7] Rocklöv J, Quam MB, Sudre B, German M, Kraemer MU, et al. Assessing seasonal risks for the introduction and mosquito-borne spread of Zika virus in Europe. eBioMedicine. 2016;**9**:250-256. DOI: 10.1016/j.ebiom.2016.06.009

[8] Dhole P, Zaidi A, Nariya HK, Sinha S, Jinesh S, Srivastava S. Host immune response to dengue virus infection: Friend or foe? Immuno. 2024;**4**:549-577. DOI: 10.3390/immuno4040033

[9] Halstead SB. Neutralization and antibody-dependent enhancement of dengue viruses. Advances in Virus Research. 2003;**60**:421-467. DOI: 10.1016/S0065-3527(03)60011-4

[10] Dejnirattisai W, Jumnainsong A, Onsirisakul N, et al. Cross-reacting antibodies enhance dengue virus infection in humans. Science. 2010;**328**(5979):745-748. DOI: 10.1126/science.1185181

[11] Rouvinski A, Guardado-Calvo P, Barba-Spaeth G, et al. Recognition determinants of broadly neutralizing human antibodies against dengue viruses. Nature. 2015;**520**(7545):109-113. DOI: 10.1038/nature14130

[12] Lai CY, Tsai WY, Lin SR, Kao CL, Hu HP, King CC, et al. Antibodies to envelope glycoprotein of dengue virus during the natural course of infection are predominantly cross-reactive and recognize epitopes containing highly conserved residues at the fusion loop of domain II. Journal of Virology. 2008;**82**(13):6631-6643. DOI: 10.1128/JVI.00316-08

[13] Crill WD, Hughes HR, Delorey MJ, Chang GJ. Humoral immune responses of dengue fever patients using epitope-specific serotype-2 virus-like particle antigens. PLoS One. 2009;**2009**(4):e4991. DOI: 10.1371/journal.pone.0004991

[14] Ripoll D, Wallqvist A, Chaudhury S. Molecular simulations reveal the role of antibody fine specificity and viral maturation state on antibody-dependent enhancement of infection in dengue virus. Frontiers in Cellular and Infection Microbiology. 2019;**9**:200. DOI: 10.3389/fcimb.2019.00200

[15] John A, Rathore A. Adaptive immune responses to primary and secondary dengue virus infections. Nature Reviews. Immunology. 2019;**19**:218-230. DOI: 10.1038/s41577-019-0123-x

[16] Ong EZ, Budigi Y, Tan HC, Robinson LN, Rowley KJ, Winnett A, et al. Preclinical evaluation of VIS513, a therapeutic antibody against dengue virus, in non-human primates. Antiviral Research. 2017;**144**:44-47. DOI: 10.1016/j.antiviral.2017.05.007

[17] Gunale B, Farinola N, Kamat CD, Poonawalla CS, Pisal SS, Dhere RM, et al. An observer-blind, randomised, placebo-controlled, phase 1, single ascending dose study of dengue monoclonal antibody in healthy adults in Australia. The Lancet. Infectious Diseases. 2024;**24**(6):639-649. DOI: 10.1016/S1473-3099(24)00030-6

[18] National Library of Medicine (U.S). Safety Study of Dengushield in Healthy Adults. Identifier Identifier NCT03883620. 2020. Available from: https://clinicaltrials.gov/study/NCT03883620

[19] National Library of Medicine (U.S). A Study to Evaluate the Safety and Efficacy of AV-1 Against Dengue Virus 3 (DENV-3) Infection. Identifier NCT06799741. 2025. Available from: https://clinicaltrials.gov/study/NCT06799741

[20] Falgout B, Markoff L. Evidence that flavivirus NS1-NS2A cleavage is mediated by a membrane-bound host protease in the endoplasmic reticulum. Journal of Virology. 1995;**69**(11):7232-7243. DOI: 10.1128/JVI.69.11.7232-7243.1995

[21] Hasan SS, Sevvana M, Kuhn RJ, Rossmann MG. Structural biology of Zika virus and other flaviviruses. Nature Structural & Molecular Biology. 2018;**25**:13-20. DOI: 10.1038/s41594-017-0010-8

[22] Modis Y, Ogata S, Clements D, Harrison SC. A ligand-binding pocket in the dengue virus envelope glycoprotein. Proceedings of the National Academy of Sciences of the United States of America. 2003;**100**(12):6986-6991. DOI: 10.1073/pnas.0832193100

[23] Mukhopadhyay S, Kuhn RJ, Rossmann MG. A structural perspective of the flavivirus life cycle. Nature Reviews. Microbiology. 2005;**3**(1):13-22. DOI: 10.1038/nrmicro1067

[24] Falconar AK. Identification of an epitope on the dengue virus membrane (M) protein defined by cross-protective monoclonal antibodies: Design of an improved epitope sequence based on common determinants present in both envelope (E and M) proteins. Archives of Virology. 1999;**144**(12):2313-2330. DOI: 10.1007/s007050050646

[25] Roehrig JT. Antigenic structure of flavivirus proteins. Advances in Virus Research. 2003;**59**:141-175. DOI: 10.1016/s0065-3527(03)59005-4

[26] Lazaro-Olan L, Mellado-Sanchez G, Garcia-Cordero J, Escobar Gutierrez A, Santos-Argumedo L, Gutierrez-Castaneda B, et al. Analysis of antibody response in human dengue patients from the Mexican coast using recombinant antigens. Vector Borne and Zoonotic Diseases. 2008;**8**:69-79. DOI: 10.1089/vbz.2007.0117

[27] Perera-Lecoin M, Meertens L, Carnec X, Amara A. Flavivirus entry receptors: An update. Viruses. 2013;**6**(1):69-88. DOI: 10.3390/v6010069

[28] Sukupolvi-Petty S, Austin SK, Engle M, Brien JD, Dowd KA, Williams KL, et al. Structure and function analysis of therapeutic monoclonal antibodies against dengue virus type 2. Journal of Virology.

2010;**84**(18):9227-9239. DOI: 10.1128/JVI.01087-10

[29] Dai L, Song J, Lu X, et al. Structures of the Zika virus envelope protein and its complex with a flavivirus broadly protective antibody. Cell Host & Microbe. 2016;**19**(5):696-704. DOI: 10.1016/j.chom.2016.04.013

[30] Brien JD, Austin SK, Sukupolvi-Petty S, et al. Genotype-specific neutralization and protection by antibodies against dengue virus type 3. Journal of Virology. 2010;**84**(20):10630-10643. DOI: 10.1128/JVI.01190-10

[31] Stettler K, Beltramello M, Espinosa DA, Graham V, Cassotta A, Bianchi S, et al. Specificity, cross-reactivity, and function of antibodies elicited by Zika virus infection. Science. 2016;**353**(6301):823-826. DOI:10.1126/science.aaf8505

[32] Dejnirattisai W, Wongwiwat W, Supasa S, et al. A globally neutralizing antibody binds to an epitope conserved across dengue virus serotypes. Nature Immunology. 2015;**16**(2):170-177. DOI: 10.1038/ni.3058

[33] Wahala WMPB, Kraus AA, Haymore LB, Accavitti-Loper MA, de Silva AM. Dengue virus neutralization by human immune sera: Role of envelope protein domain III-reactive antibody. Virology. 2009;**392**(1):103-113. DOI: 10.1016/j.virol.2009.06.037

[34] Fernandez E, Dejnirattisai W, Cao B, Scheaffer SM, Supasa P, Wongwiwat W, et al. Human antibodies to the dengue virus E-dimer epitope have therapeutic activity against Zika virus infection. Nature Immunology. 2017;**18**(11):1261-1269. DOI: 10.1038/ni.3849

[35] Wang R, Lu J, Chen L, Yu Y. Yang Z. A human bispecific neutralization antibody against four serotypes of dengue virus. Virology. 2021;**558**:49-56. DOI: 10.1016/j.virol.2021.01.009

[36] De Alwis R, Smith SA, Olivarez NP, Messer WB, Huynh JP, Wahala WMPB, et al. Identification of human neutralizing antibodies that bind to complex epitopes on dengue virions. Proceedings of the National Academy of Sciences of the United States of America. 2012;**109**(19):7439-7444. DOI: 10.1073/pnas.1200566109

[37] Li L, Lok SM, Yu IM, et al. The flavivirus precursor membrane-envelope protein complex: Structure and maturation. Science. 2008;**319**(5871):1830-1834. DOI: 10.1126/science.1153263

[38] Lu H, Xu XF, Gao N, Fan DJ, Wang J, An J. Preliminary evaluation of DNA vaccine candidates encoding dengue-2 prM/E and NS1: Their immunity and protective efficacy in mice. Molecular Immunology. 2013;**54**(2):109-114. DOI: 10.1016/j.molimm.2012.11.007

[39] Beltramello M, Williams KL, Simmons CP, Macagno A, Simonelli L, Quyen NT, et al. The human immune response to dengue virus is dominated by highly cross-reactive antibodies endowed with neutralizing and enhancing activity. Cell Host & Microbe. 2010;**8**(3):271-283. DOI: 10.1016/j.chom.2010.08.007

[40] Smith SA, Zhou Y, Olivarez NP, Broadwater AH, de Silva AM, Crowe JE. Persistence of circulating memory B cell clones with potential for dengue virus disease enhancement for decades following infection. Journal of Virology. 2012;**86**(5):2665-2675. DOI: 10.1128/JVI.06335-11

[41] Chan AH, Tan HC, Chow AY, Lim AP, Lok SM, Moreland NJ, et al. A human PrM antibody that recognizes a novel cryptic epitope on

dengue E glycoprotein. PLoS One. 2012;7(4):e33451. DOI: 10.1371/journal.pone.0033451

[42] Flamand M, Megret F, Mathieu M, Lepault J, Rey FA, Deubel V. Dengue virus type 1 nonstructural glycoprotein NS1 is secreted from mammalian cells as a soluble hexamer in a glycosylation-dependent fashion. Journal of Virology. 1999;**73**(7):6104-6110. DOI: 10.1128/JVI.73.7.6104-6110

[43] Henchal EA, Henchal LS, Schlesinger JJ. Synergistic interactions of anti NS1 monoclonal antibodies protect passively immunized mice from lethal challenge with dengue 2 virus. The Journal of General Virology. 1988;**69**:2101-2107. DOI: 10.1099/0022-1317-69-8-2101

[44] Glasner DR, Ratnasiri K, Puerta-Guardo H, Espinosa DA, Beatty PR, Harris E. Dengue virus NS1 cytokine-independent vascular leak is dependent on endothelial glycocalyx components. PLoS Pathogens. 2017;**13**(11):e1006673. DOI: 10.1371/journal.ppat.1006673

[45] Falconar AK. The dengue virus nonstructural-1 protein (NS1) generates antibodies to common epitopes on human blood clotting, integrin/adhesin proteins and binds to human endothelial cells: Potential implications in haemorrhagic fever pathogenesis. Archives of Virology. 1997;**142**(5):897-916. DOI: 10.1007/s007050050127

[46] Kanlaya R, Pattanakitsakul SN, Sinchaikul S, Chen ST, Thongboonkerd V. Vimentin interacts with heterogeneous nuclear ribonucleoproteins and dengue nonstructural protein 1 and is important for viral replication and release. Molecular BioSystems. 2010;**6**(5):795-806. DOI: 10.1039/b923864f

[47] Chuang YC, Wang SY, Lin YS, Chen HR, Yeh TM. Re-evaluation of the pathogenic roles of nonstructural protein 1 and its antibodies during dengue virus infection. Journal of Biomedical Science. 2013;**20**(1):42. DOI: 10.1186/1423-0127-20-42

[48] Muller DA, Young PR. The flavivirus NS1 protein: Molecular and structural biology, immunology, role in pathogenesis and application as a diagnostic biomarker. Antiviral Research. 2013;**98**(2):192-208. DOI: 10.1016/j.antiviral.2013.03.008

[49] Cedillo-Barrón L, García-Cordero J, Bustos-Arriaga J, León-Juárez M, Gutiérrez-Castañeda B. Antibody response to dengue virus. Microbes and Infection. 2014;**16**:711-720. DOI: 10.1016/j.micinf.2014.07.011

[50] Imrie A, Meeks J, Gurary A, Sukhbaatar M, Truong TT, Cropp CB, et al. Antibody to dengue 1 detected more than 60 years after infection. Viral Immunology. 2007;**20**:672-675. DOI: 10.1089/vim.2007.0050

[51] Andrade P, Gimblet-Ochieng C, Modirian F, et al. Impact of pre-existing dengue immunity on human antibody and memory B cell responses to Zika. Nature Communications. 2019;**10**:938. DOI: 10.1038/s41467-019-08845-3

[52] Rothman AL. Immunity to dengue virus: A tale of original antigenic sin and tropical cytokine storms. Nature Reviews. Immunology. 2011;**11**(8):532-543. DOI: 10.1038/nri3014

[53] Halstead SB, O'Rourke EJ. Antibody-enhanced dengue virus infection in primate leukocytes. Nature. 1977;**265**(5596):739-741. DOI: 10.1038/265739a0

[54] Kliks SC, Nisalak A, Brandt WE, Wahl L, Burke DS. Antibody-dependent

enhancement of dengue virus growth in human monocytes as a risk factor for dengue hemorrhagic fever. The American Journal of Tropical Medicine and Hygiene. 1989;**40**(4):444-451. DOI: 10.4269/ajtmh.1989.40.444

[55] Sangkawibha N, Rojanasuphot S, Ahandrik S, Viriyapongse S, Jatanasen S, et al. Risk factors in dengue shock syndrome: A prospective epidemiologic study in Rayong, Thailand. I. The 1980 outbreak. American Journal of Epidemiology. 1984;**120**(5):653-669. DOI: 10.1093/oxfordjournals.aje.a113932

[56] Guzman MG, Harris E. Dengue. Lancet. 2015;**385**(9966):453-465. DOI: 10.1016/S0140-6736(14)60572-9

[57] Halstead SB. Pathogenesis of dengue: Challenges to molecular biology. Science. 1988;**239**(4839):476-481. DOI: 10.1126/science.3277268

[58] Mancardi D, Daëron M. Fc Receptors in immune responses. In: Caplan M, editor. Reference Module in Biomedical Sciences. 1st ed. Elsevier; 2014. p.1-14. DOI: 10.1016/B978-0-12-801238-3.00119-7

[59] Sarker A, Dhama N, Gupta RD. Dengue virus neutralizing antibody: A review of targets, cross-reactivity, and antibody-dependent enhancement. Frontiers in Immunology. 2023;**14**:1200195. DOI: 10.3389/fimmu.2023.1200195

[60] Chareonsirisuthigul T, Kalayanarooj S, Ubol S. Dengue virus (DENV) antibody- dependent enhancement of infection upregulates the production of anti-inflammatory cytokines, but suppresses anti-DENV free radical and pro-inflammatory cytokine production, in THP-1 cells. The Journal of General Virology. 2007;**88**(Pt 2):365-375. DOI: 10.1099/vir.0.82537-0

[61] Modhiran N, Kalayanarooj S, Ubol S. Subversion of innate defenses by the interplay between DENV and pre-existing enhancing antibodies: TLRs signaling collapse. PLoS Neglected Tropical Diseases. 2010;**4**(12):e924. DOI: 10.1371/journal.pntd.0000924

[62] Halstead SB. Dengue antibody-dependent enhancement: Knowns and unknowns. Microbiology Spectrum. 2014;**2**(6):1-18. DOI: 10.1128/microbiolspec.AID-0022-2014

[63] Bhatt P, Sabeena SP, Varma M, Arunkumar G. Current understanding of the pathogenesis of dengue virus infection. Current Microbiology. 2021;**78**(1):17-32. DOI: 10.1007/s00284-020-02284-w

[64] Lin CF, Wan SW, Chen MC, et al. Liver injury caused by antibodies against dengue virus nonstructural protein 1 in a murine model. Laboratory Investigation. 2008;**88**(10):1079-1089. DOI: 10.1038/labinvest.2008.70

[65] Avirutnan P, Zhang L, Punyadee N, et al. Secreted NS1 of dengue virus attaches to the surface of cells via interactions with heparan sulfate and chondroitin sulfate E. PLoS Pathogens. 2006;**3**(11):e183. DOI: 10.1371/journal.ppat.0030183

[66] Priyamvada L, Quicke KM, Hudson WH, Onlamoon N, Sewatanon J, Edupuganti S, et al. Human antibody responses after dengue virus infection are highly cross-reactive to Zika virus. Proceedings of the National Academy of Sciences of the United States of America. 2016;**113**(28):7852-7857. DOI: 10.1073/pnas.1607931113

[67] Dejnirattisai W, Supasa P, Wongwiwat W, et al. Dengue virus sero-cross-reactivity drives antibody-dependent enhancement of infection

with zika virus. Nature Immunology. 2016;**17**:1102-1108. DOI: 10.1038/ni.3515

[68] Bardina SV, Bunduc P, Tripathi S, Duehr J, Frere JJ, et al. Enhancement of Zika virus pathogenesis by preexisting antiflavivirus immunity. Science. 2017;**356**(6334):175-180. DOI: 10.1126/science.aal4365

[69] Goncalvez AP, Engle RE, St Claire M, Purcell RH, Lai CJ. Monoclonal antibody- mediated enhancement of dengue virus infection in vitro and in vivo and strategies for prevention. Proceedings of the National Academy of Sciences of the United States of America. 2007;**104**(22):9422-9427. DOI: 10.1073/pnas.0703498104

[70] Nandi S, Kwong AT, Holtz BR, Erwin RL, Marcel S, McDonald KA. Techno-economic analysis of a transient plant-based platform for monoclonal antibody production. MAbs. 2016;**8**(8):1456-1466. DOI: 10.1080/19420862.2016.1227901

[71] Injampa S, Muenngern N, Pipattanaboon C, Benjathummarak S, Boonha K, Hananantachai H, et al. Generation and characterization of cross neutralizing human monoclonal antibody against 4 serotypes of dengue virus without enhancing activity. PeerJ. 2017;**5**:e4021. DOI: 10.7717/peerj.4021

[72] Williams KL, Sukupolvi-Petty S, Beltramello M, et al. Therapeutic efficacy of antibodies lacking Fcγ receptor binding against dengue virus. Nature Medicine. 2013;**19**(4):456-460. DOI: 10.1371/journal.ppat.1003157

[73] Shi X, Deng Y, Wang H, Ji G, Tan W, Jiang T, et al. A bispecific antibody effectively neutralizes all four serotypes of dengue virus by simultaneous blocking virus attachment and fusion. MAbs. 2016;**8**(3):574-584. DOI: 10.1080/19420862.2016.1148850

[74] Pinto AK, Brien JD, Lam CY, Johnson S, Chiang C, Hiscott J, et al. Defining new therapeutics using a more immunocompetent mouse model of antibody-enhanced dengue virus infection. MBio. 2015;**6**(5):e01316-e01315. DOI: 10.1128/mBio.01316-15

[75] Brien JD, Sukupolvi-Petty S, Williams KL, Lam CY, Schmid MA, Johnson S, et al. Protection by immunoglobulin dual-affinity retargeting antibodies against dengue virus. Journal of Virology. 2013;**87**:7747-7753. DOI: 10.1128/JVI.00327-13

[76] Guy B, Briand O, Lang J, Saville M, Jackson N. Development of the Sanofi Pasteur tetravalent dengue vaccine: One more step forward. Vaccine. 2015;**33**(50):7100-7111. DOI: 10.1016/j.vaccine.2015.09.108

[77] Hadinegoro SR, Arredondo-García JL, Capeding MR, Deseda C, Chotpitayasunondh T, et al. Efficacy and long-term safety of a dengue vaccine in regions of endemic disease. The New England Journal of Medicine. 2015;**373**(13):1195-1206. DOI: 10.1056/NEJMoa1506223

[78] Sridhar S, Luedtke A, Langevin E, Zhu M, Bonaparte M, Machabert T, et al. Effect of dengue serostatus on dengue vaccine safety and efficacy. The New England Journal of Medicine. 2018;**379**(4):327-340. DOI: 10.1056/NEJMoa1800820

[79] Halstead SB, Katzelnick LC, Russell PK, Markoff L, Aguiar M, Dans LR, et al. Ethics of a partially effective dengue vaccine: Lessons from the Philippines. Vaccine. 2020;**38**(35):5572-5576. DOI: 10.1016/j.vaccine.2020.06.079

[80] Osorio JE, Huang CY, Kinney RM, Stinchcomb DT. Development of DENVax: A chimeric dengue-2 PDK-53-based tetravalent vaccine for protection against dengue fever. Vaccine. 2011;**29**(42):7251-7260. DOI: 10.1016/j.vaccine.2011.07.020

[81] Biswal S, Reynales H, Saez-Llorens X, Lopez P, Borja-Tabora C, et al. Efficacy of a tetravalent dengue vaccine in healthy children and adolescents. The New England Journal of Medicine. 2019;**381**(21):2009-2019. DOI: 10.1056/NEJMoa1903869

[82] Whitehead SS. Development of TV003/TV005, a single dose, highly immunogenic live attenuated dengue vaccine; what makes this vaccine different from the Sanofi-Pasteur CYD™ vaccine? Expert Review of Vaccines. 2016;**15**(4):509-517. DOI: 10.1586/14760584.2016.1115727

[83] Durbin AP, Kirkpatrick BD, Pierce KK, Elwood D, Larsson CJ, et al. A single dose of any of four different live attenuated tetravalent dengue vaccines is safe and immunogenic in flavivirus-naive adults: A randomized, double-blind clinical trial. The Journal of Infectious Diseases. 2013;**207**(6):957-965. DOI: 10.1093/infdis/jis936

[84] Akter R, Tasneem F, Das S, Soma MA, Georgakopoulos-Soares I, et al. Approaches of dengue control: Vaccine strategies and future aspects. Frontiers in Immunology. 2024;**29**(15):1362780. DOI: 10.3389/fimmu.2024.1362780

[85] Yan K, Mao L, Lan J, Xiao Z. Advancements in dengue vaccines: A historical overview and prospects for following next-generation candidates. Journal of Microbiology. 2025;**63**(2):e2410018. DOI: 10.71150/jm.2410018

[86] Putnak RJ, Coller BA, Voss G, Vaughn DW, Clements D, et al. An evaluation of dengue type-2 inactivated, recombinant subunit, and live-attenuated vaccine candidates in the rhesus macaque model. Vaccine. 2005;**23**(35):4442-4452. DOI: 10.1016/j.vaccine.2005.03.042

[87] Schmidt AC, Lin L, Martinez LJ, Ruck RC, Eckels KH, et al. Phase 1 randomized study of a tetravalent dengue purified inactivated vaccine in healthy adults in the United States. The American Journal of Tropical Medicine and Hygiene. 2017;**96**(6):1325-1337. DOI: 10.4269/ajtmh.16-0634

[88] Friberg H, Martinez LJ, Lin L, Blaylock JM, De La Barrera RA, et al. Cell-mediated immunity generated in response to a purified inactivated vaccine for dengue virus type 1. mSphere. 2020;**5**(1):e00671-e00619. DOI: 10.1128/mSphere.00671-19

[89] Wu SJ, Ewing D, Sundaram AK, Chen HW, Liang Z, et al. Enhanced immunogenicity of inactivated dengue vaccines by novel polysaccharide-based adjuvants in mice. Microorganisms. 2022;**10**(5):1034. DOI: 10.3390/microorganisms10051034

[90] Mani S, Tripathi L, Raut R, Tyagi P, Arora U, et al. Pichia pastoris-expressed dengue 2 envelope forms virus-like particles without pre-membrane protein and induces high titer neutralizing antibodies. PLoS One. 2013;**8**(5):e64595. DOI: 10.1371/journal.pone.0064595

[91] Tripathi NK, Shrivastava A. Recent developments in recombinant protein-based dengue vaccines. Frontiers in Immunology. 2018;**9**:1919. DOI: 10.3389/fimmu.2018.01919

[92] Govindarajan D, Meschino S, Guan L, Clements DE, ter Meulen JH,

et al. Preclinical development of a dengue tetravalent recombinant subunit vaccine: Immunogenicity and protective efficacy in nonhuman primates. Vaccine. 2015;**33**(33):4105-4116. DOI: 10.1016/j.vaccine.2015.06.067

[93] Ramasamy V, Arora U, Shukla R, Poddar A, Shanmugam RK, White LJ, et al. A tetravalent virus-like particle vaccine designed to display domain III of dengue envelope proteins induces multi-serotype neutralizing antibodies in mice and macaques which confer protection against antibody dependent enhancement in AG129 mice. PLoS Neglected Tropical Diseases. 2018;**12**(1):e0006191. DOI: 10.1371/journal.pntd.0006191

[94] Hsiung K-C, Chiang H-J, Reinig S, Shih S-R. Vaccine strategies against RNA viruses: Current advances and future directions. Vaccine. 2024;**12**(12):1345. DOI: 10.3390/vaccines12121345

[95] Khanam S, Pilankatta R, Khanna N, Swaminathan S. An adenovirus type 5 (AdV5) vector encoding an envelope domain III-based tetravalent antigen elicits immune responses against all four dengue viruses in the presence of prior AdV5 immunity. Vaccine. 2009;**27**(43):6011-6021. DOI: 10.1016/j.vaccine.2009.07.073

[96] Saxena M, Van TTH, Baird FJ, Coloe PJ, Smooker PM. Pre-existing immunity against vaccine vectors—Friend or foe? Microbiology. 2013;**159**(Pt 1):1-11. DOI: 10.1099/mic.0.049601-0

[97] Khoshnood S, Ghanavati R, Shirani M, Ghahramanpour H, Sholeh M, Shariati A, et al. Viral vector and nucleic acid vaccines against COVID-19: A narrative review. Frontiers in Microbiology. 2022;**31**(13):984536. DOI: 10.3389/fmicb.2022.984536

[98] Leitner WW, Ying H, Restifo NP. DNA and RNA-based vaccines: Principles, progress and prospects. Vaccine. 1999;**18**(9-10):765-777. DOI: 10.1016/s0264-410x(99)00271-6

[99] Manoj S, Babiuk LA, van Drunen Littel-van den Hurk S. Approaches to enhance the efficacy of DNA vaccines. Critical Reviews in Clinical Laboratory Sciences. 2004;**41**(1):1-39. DOI: 10.1080/10408360490269251

[100] Beckett CG, Tjaden J, Burgess T, Danko JR, Tamminga C, et al. Evaluation of a prototype dengue-1 DNA vaccine in a phase 1 clinical trial. Vaccine. 2011;**29**(5):960-968. DOI: 10.1016/j.vaccine.2010.11.050

[101] Fang E, Liu X, Li M, Zhang Z, Song L, et al. Advances in COVID-19 mRNA vaccine development. Signal Transduction and Targeted Therapy. 2022;**7**(1):94. DOI: 10.1038/s41392-022-00950-y

[102] Zhang M, Sun J, Li M, Jin X. Modified mRNA-LNP vaccines confer protection against experimental DENV-2 infection in mice. Molecular Therapy - Methods & Clinical Development. 2020;**18**:702-712. DOI: 10.1016/j.omtm.2020.07.013

[103] He L, Sun W, Yang L, Liu WL. A multiple-target mRNA-LNP vaccine induces protective immunity against experimental multi-serotype DENV in mice. Virologica Sinica. 2022;**37**(5):746-757. DOI: 10.1016/j.virs.2022.07.003

Chapter 4

# Vaccines and Immune Response to *Orthoflavivirus*: Challenges and Advances

*Samantha Climaco-Arvizu and Mara Gutiérrez-Sánchez*


## Abstract

*Orthoflavivirus*, a genus of positive-sense, single-stranded RNA viruses belonging to the family *Flaviviridae*, infects vertebrates and is transmitted by arthropods (arboviruses). These include pathogens such as dengue, yellow fever, Zika, and Japanese encephalitis viruses. These are clinically relevant viruses that can cause symptoms ranging from mild to severe forms, with hemorrhagic and neurological involvement, posing significant threats to human health worldwide. The immune response to *Orthoflavivirus* involves a complex interaction between innate and adaptive immunity, with T cells and neutralizing antibodies playing key roles. Cross-reactivity between serotypes can lead to ineffective viral control and complicate both immunopathology and vaccine development. It should be noted that there are still no approved antiviral treatments against *Orthoflavivirus*. -Therefore, strategies have been implemented to control these infections, such as vector management and vaccines. Some have already been approved and have been instrumental in significantly reducing these infections, as was the case with yellow fever, whose vaccine served as the basis for the development of current vaccines. Despite these achievements, vaccine efficacy remains variable. This chapter will focus on providing a comprehensive understanding of the advances related to *Orthoflavivirus,* including their immunological characteristics, immune responses, vaccination strategies, and current challenges.


**Keywords:** *Orthoflavivirus*, vaccines, immunological characteristics of *Orthoflavivirus*, immune response, vaccine development, vaccine efficacy, universal vaccine

## 1. Introduction

The genus *Orthoflavivirus* belongs to the *Flaviviridae* family and consists of arthropod-borne viruses (arboviruses) transmitted by vectors such as mosquitoes and ticks [1]. Major human pathogens include dengue virus (DENV), Zika virus (ZIKV), yellow fever virus (YFV), and West Nile virus (WNV), which can cause a wide range of clinical symptoms, from mild (asymptomatic infection or self-limiting febrile episodes) to severe and life-threatening disease (systemic disease involving hemorrhage, shock syndrome, neurological complications, and congenital defects). These viruses are also associated with severe outbreaks and pose a significant global health threat [2, 3]. Currently, there are no antiviral treatments in clinical use for *Orthoflavivirus* infections. Although a few vaccines have been approved for human use, their effectiveness has been variable [4]. Therefore, the development of effective disease-control strategies, particularly vaccines, remains crucial.

### 1.1 Global burden and epidemiological landscape of *Orthoflavivirus* infections

*Orthoflaviviridae* are important pathogens that infect humans and are responsible for widespread morbidity and mortality worldwide. Dengue (DEN) is considered the most important arboviral disease, causing the highest rates of morbidity and mortality among members of the *Orthoflavivirus* genus [5]. DEN is endemic in more than 100 countries, with 90 countries known to have active DEN transmission, and 161 countries or territories included in the WHO surveillance system. According to the WHO, more than 7.6 million cases of DEN were reported in 2024, including 3.4 million confirmed cases, over 16,000 severe cases, and more than 3000 deaths. Many endemic countries do not have strong detection and reporting mechanisms, so the true global burden of DEN is underestimated [6]. As of 2024, the number of DEN cases in the Region of the Americas has already exceeded seven million, with all four DEN serotypes (DENV-1, DENV-2, DENV-3, and DENV-4) detected throughout the region [6]. Africa has reported a total of 32,925 DEN cases, including 14,095 confirmed cases, 1051 severe cases, and 57 deaths [6]. In the Western Pacific, Malaysia was the most affected, reporting 50,650 cases and 39 deaths, followed by Vietnam with 16,111 cases and one death [6]. Globally, the health and economic burden of DEN is significant, with an average annual cost of US$76.5 billion between 1975 and 2020, and annual costs ranging between US$1.7 billion and US$17.5 billion [6, 7].

Despite its long history and sporadic infection cases, the ZIKV was not declared a global public health emergency by the WHO until 2016, following a major outbreak in Brazil that led to widespread concern and international response [8]. A modeling study based on African data sources estimated the burden of yellow fever (YF) in 2013 at 84,000–170,000 severe cases and 29,000–60,000 deaths [9]. In 2024, 61 human cases of YF were confirmed in the Region of the Americas, including 30 deaths. Humans primarily acquire DENV, ZIKV, and YFV through horizontal transmission *via* the bite of infected female mosquitoes, primarily *Aedes aegypti* (Diptera, Culicidae) [10, 11]. Additionally, *Aedes albopictus* can transmit DENV and ZIKV [11]. In contrast, WNV is transmitted by mosquitoes of a different genus, with *Culex* species being the most common vector [12]. Currently, nine distinct lineages of WNV (WNV-1 through WNV-9) have been identified worldwide, with WNV-1 and WNV-2 being the most detected strains in human cases across several continents [13]. WNV is present in Africa, Europe, Asia, North America, Australia, and the Middle East [14].

### 1.2 Clinical manifestations of *Orthoflavivirus* infections

After an infectious mosquito bite, symptoms of *Orthoflavivirus* infection typically develop within 7 days and last between 3 and 14 days. The four serotypes of DENV can cause a range of clinical diseases in humans, from self-limited DEN fever to a life-threatening syndrome known as severe DEN [2]. The clinical presentation of DENV infection has been classified into three phases: A febrile phase, a critical phase, and a recovery phase; however, the critical phase is not seen in all categories of infection [15]. The febrile phase of DENV infection usually lasts between 3 and 7 days. Most patients recover without complications. This phase is typically characterized by high fever (>38.5°C), headache, vomiting, myalgia, arthralgia, retro-orbital pain, hemorrhagic manifestations, and, in some cases, a macular rash [16]. Most infections that progress to a critical phase result from secondary DENV infections that occur more than 18 months after a resolved first infection. In addition, severe DENV infection may occur after primary infection in individuals with significant medical comorbidities [17, 18]. The critical phase of DEN begins at defervescence and typically lasts 24–48 hours [15]. The progression of DEN fever to DEN hemorrhagic fever (DHF) might suggest that hemorrhage is the primary manifestation of severe DEN. However, plasma leakage leading to intravascular volume depletion and potentially shock, known as DEN shock syndrome (DSS), is the most specific feature of severe DEN and the focus of clinical management guidelines and algorithms [15, 19]. Patients may appear to be well despite early signs of shock. However, once hypotension develops, systolic blood pressure rapidly declines, and irreversible shock and death may ensue despite resuscitation [19]. Additional manifestations of DENV infection (typically occurring in the critical phase or later) may include liver failure, central nervous system involvement, myocardial dysfunction, acute kidney injury, and others [20, 21].

ZIKV, in addition to being vector-transmitted, can also spread from mother to fetus during pregnancy, through sexual contact, transfusion of blood and blood products, and possibly through organ transplantation [22]. ZIKV is strongly associated with microcephaly in infants born to infected mothers and Guillain-Barré syndrome in adults. Each cell type infected by ZIKV—neuronal and non-neuronal—exhibits its own characteristic changes in cell physiology and has different effects on disease [23]. In YFV infection, symptomatic individuals present with fever, severe back pain, and headache, and in most cases, the symptoms disappear after 3–4 days; however, a small percentage of patients enter a second, more toxic phase within 24 hours of recovering from the initial symptoms [24]. High fever returns, and multiple body systems are affected, usually the liver and kidneys. During this phase, people are likely to develop jaundice (yellowing of the skin and eyes, hence the name "YF"), dark urine, and abdominal pain with vomiting. Half of patients who enter the toxic phase die within 7–10 days [24]. WNV typically presents with myalgia, malaise, and low-grade fever, accompanied by headache, eye pain, vomiting, and anorexia. Up to 50% of cases may develop a maculopapular rash on the trunk upon defervescence. In rare instances, WNV can lead to neurological complications. These patients often exhibit signs of encephalitis and/or meningitis, which rapidly progress and necessitate intensive care. Despite aggressive supportive management, neuroinvasive WNV carries a high mortality rate [14].

### 1.3 Control strategies for the prevention of *Orthoflavivirus* infections

Effective global strategies to detect, treat, and limit *Orthoflavivirus* transmission are crucial to reducing its public health impact and achieving sustainable control. Equally important is the development of effective vaccines and the implementation of vaccination programs in regions with high case prevalence, as well as ensuring equitable access to these treatments in areas where economic and geographic challenges frequently limit availability [25]. Since its arrival in the United States in 1999, WNV has become the most common mosquito-borne virus in North America. Despite over two decades of intensive research, there are still no approved vaccines or specific treatments for humans, and there remains an urgent need to understand the pathogenesis of WNV and to develop specific therapeutics and vaccines [25]. Research efforts continue as the pathogenesis and immune evasion mechanisms of ZIKV remain incompletely understood. Currently, no specific vaccines or antiviral drugs have been approved for ZIKV; however, several candidates are progressing through clinical trials, offering promise for future treatment and prevention strategies [8].

## 2. General characteristics of *Orthoflavivirus*

### 2.1 Structural organization and life cycle of *Orthoflavivirus*

The *Orthoflavivirus* virion is a small spherical particle of ~50 nm in diameter, containing a single positive-stranded RNA genome of ~11 kb in length, packaged by a viral capsid protein arranged in icosahedral symmetry and enveloped in a host-derived lipid membrane [26, 27]. The RNA genome contains a single open reading

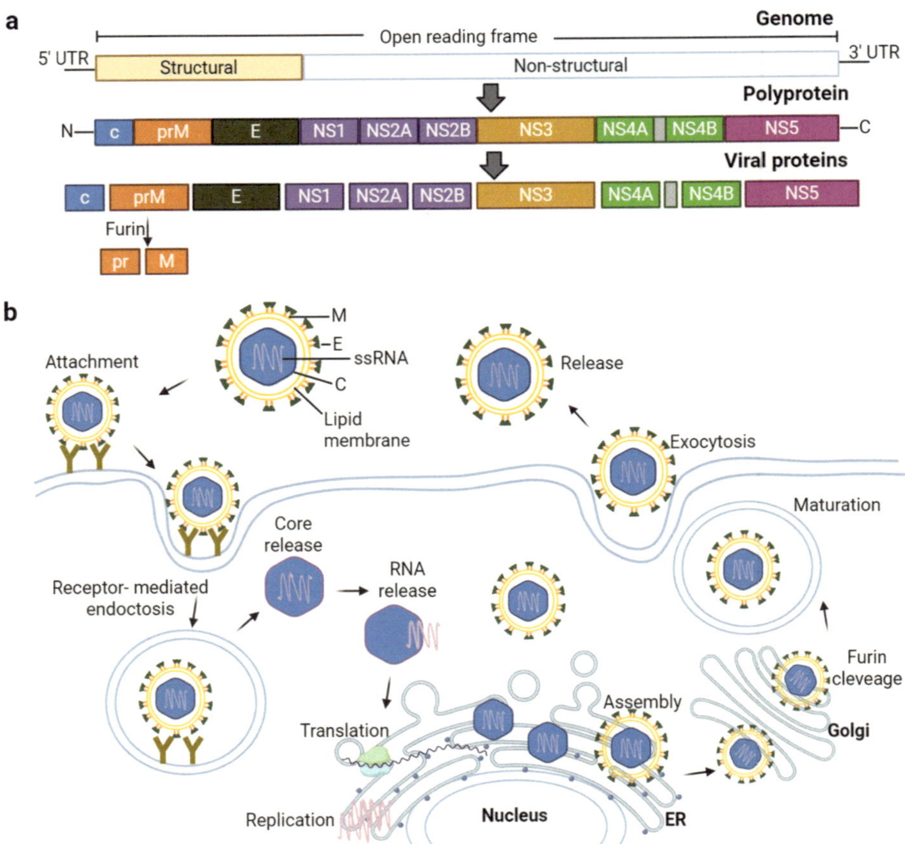


**Figure 1.**
Orthoflavivirus *structural organization and life cycle. (a)* Orthoflavivirus *genome organization and processing of the polypeptide into mature viral proteins. (b) Viral particles bind to receptors on the surface of the host cell membrane and enter the cell by endocytosis, forming an early endosome. Acidification of the endosome induces conformational changes that release the nucleocapsid into the cytoplasm. The viral genome is then released into the cytoplasm, where it translocate into the endoplasmic reticulum (ER) and is directly translated into a polyprotein, which is cleaved and processed by viral and host proteases. In addition, replication complexes replicate viral RNA through a negative-stranded RNA intermediate to produce positive-stranded RNA. Virus assembly occurs on the surface of the ER. Immature, noninfectious virus particles are transported to the trans-Golgi network (TGN). The host protease furin cleaves prM to M, forming mature virus particles that are released by exocytosis, completing the replication cycle.*

frame (ORF) flanked by a 5′-untranslated region (UTR) and a 3′-UTR, encoding three structural proteins (the capsid (C), precursor membrane (prM) and envelope glycoprotein (E)), which form the viral particle and seven nonstructural (NS) proteins (NS1, NS2A, NS2B, NS3, NS4A, NS4B, and NS5) that play several roles in viral life cycle and in initiating host innate immunity [26, 28]. Dendritic cells are a common initial target for *Orthoflaviviruses*. Upon infection, they migrate to lymphoid organs, promoting viral replication and enabling the spreading into the circulation and internal organs [29]. The *Orthoflavivirus* viral cycle begins with the attachment to the host cell surface, followed by entry *via* receptor-mediated endocytosis, and then virions are transported into endosomes, where acidification of the endosomal vesicle triggers conformational changes in the virion, leading to fusion of the viral membrane with the endosomal membrane and particle disassembly [29]. This process results in the release of viral genomic RNA into the cell cytoplasm, where it is translated by the host ribosome machinery into a single polyprotein. The polyprotein is embedded in the endoplasmic reticulum (ER) and is cleaved by host and viral proteins to produce three structural proteins and at least seven nonstructural proteins. Genome replication and assembly of immature viral particles occur in the ER, where they bud into the lumen of the ER. Subsequently, noninfectious viral particles are transported into the Golgi apparatus for maturation by the host protease furin, resulting in mature, infectious particles. Subviral particles (noninfectious) are also produced in the ER and processed by furin. Finally, mature virions and subviral particles are released from the cell by exocytosis [30–32]. The structural organization and life cycle of *Orthoflavivirus* are shown in **Figure 1**.

## 3. Immunological characteristics of *Orthoflavivirus*

### 3.1 Innate immunity to *Orthoflavivirus*

During *Orthoflavivirus* infection, host cell pattern-recognition receptors (PRRs) located in endosomal compartments and the cytoplasm recognize and respond to viral RNA and virus-infected cells [33]. The innate immune system is the first line of defense against invading pathogens, with a rapid, nonspecific response to control infection. It also plays a key role in the establishment of adaptive immunity, which is pathogen-specific and provides long-lasting immunological memory [34]. Two events are required to trigger an effective immune response: (1) the detection of invading viral molecules by PRRs, and (2) further activation of protein signaling cascades by PRRs that regulate the synthesis and secretion of type I interferons (IFNs). PRRs recognize *Orthoflavivirus* pathogen-associated molecular patterns (PAMPs), such as viral RNA, through toll-like receptors (TLRs), retinoic acid-inducible gene I (RIG-I)-like receptors (RLRs), and cyclic guanosine monophosphate (GMP)–adenosine monophosphate (AMP) synthase (cGAS) (**Figure 2**) [35].

The TLR pathway follows two distinct routes, determined by the presence or absence of the myeloid differentiation primary response protein 88 (MyD88), with the MyD88-dependent pathway leading to nuclear factor-kappa B (NF-κB) and mitogen-activated protein kinase (MAPK) activation. All TLRs, except TLR3, trigger this MyD88-dependent pathway, while TLR3 uses the MyD88-independent toll/interleukin-1 receptor/resistance protein (TIR) domain containing adapter inducing interferon-β (TRIF) pathway to promote type I interferon (IFN) and inflammatory cytokine production [36]. TLR3 and TLR7 are located in the endosomes, where they

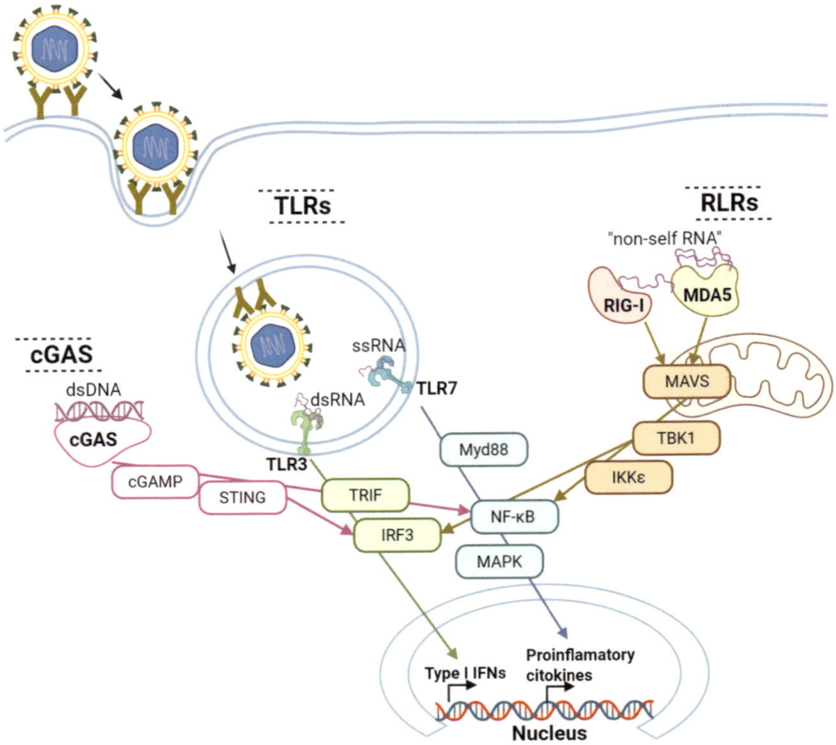


**Figure 2.**
*Innate immune pathways in* Orthoflavivirus *infection. In toll-like receptors (TLRs) signaling, TLR3 and TLR7 are located in the endosomes and recognize double-stranded RNA (dsRNA). TLR3 utilizes the toll/interleukin-1 receptor/resistance protein (TIR) domain containing adapter inducing interferon-β (TRIF) pathway to promote type I interferon (IFN) and inflammatory cytokine production, while TLR7 follows the myeloid differentiation primary response protein 88 (MyD88)-dependent pathway leading to nuclear factor-kappa B (NF-κB) and mitogen-activated protein kinase (MAPK) activation. The retinoic acid-inducible gene I (RIG-I)-like receptors (RLRs) consist of RIG-I and melanoma differentiation-associated gene 5 protein (MDA5). These RLRs recognize nonself RNA, RIG-I/MDA5 recruits mitochondrial antiviral-signaling protein (MAVS) to inhibitor of nuclear factor kappa-B kinase ε (IKKε) and TRAF family member-associated NF-κB activator (TANK) binding kinase 1 (TBK1), activating interferon regulatory factor 3 (IRF3) and NF-κB to induce type I IFN and inflammatory cytokine production. Cyclic guanosine monophosphate (GMP)–adenosine monophosphate (AMP) synthase (cGAS) binds to double-stranded DNA (dsDNA), then forms cyclic GMP-AMP (cGAMP) that activates stimulator of interferon genes (STING), then activates IRF3 and NF-κB.*

recognize double-stranded RNA (dsRNA) transiently generated during DENV, ZIKV, YFV, and WNV life cycles [37–41]. TLR3 efficiently suppresses DENV replication by upregulating IFN-α, IFN-β, and type III IFN (IL-28A/B), resulting in a significant decrease in viral copy number. Knockout of TLR3 in macrophages increased susceptibility to DENV infection, while knockdown of TLR3 in skin fibroblasts increased ZIKV viral RNA [42, 43]. TLR3 also induces type I IFN during WNV and ZIKV infections; however, the NS1 protein of WNV can suppress this pathway by inhibiting the interferon regulatory factor 3 (IRF3) transcription and activating NF-κB cytokine transcription [44]. It has been shown that WNV-infected TLR7-deficient mice had reduced IL-23, along with an increased viremia and susceptibility to lethal WNV infection [45].

The RLRs involved in the immune response against *Orthoflavivirus* consist of RIG-I and melanoma differentiation-associated protein 5 (MDA5). Once these RLRs recognize nonself RNA, RIG-I/MDA5 recruits the adapter mitochondrial antiviral-signaling protein (MAVS) to activate the noncanonical IKK-related kinases, inhibitor of nuclear factor kappa-B kinase ε (IKKε) and TRAF family member-associated NF-κB activator (TANK) binding kinase 1 (TBK1), which subsequently trigger IRF3 and NF-κB activation, inducing type I IFN and immune factors [46, 47]. DENV2, WNV, YFV, and ZIKV prM proteins significantly inhibit type I IFN production by interacting with MDA5 and/or MAVS [48]. Additionally, the phosphatases PP1α and PP1γ are essential activators of the signal transduction of both RIG-I and MDA5-mediated antiviral type I IFN production, thereby suppressing DENV replication [49].

cGAS binds to double-stranded DNA (ds-DNA) and catalyzes the formation of cyclic dinucleotides, which then generate cyclic GMP-AMP (cGAMP) that activates the stimulator of interferon genes (STING) to trigger type I IFN production. The cGAS/STING pathway is one of the major cytosolic DNA-sensing systems implicated in *Orthoflavivirus* restriction. IL-1β-induced activation of IRF3 is dependent on STING, which is triggered by cGAS recognition of cytosolic mitochondrial DNA (mtDNA) that can be released by *Orthoflavivirus* infection [50]. ZIKV NS1 stabilizes caspase-1 by blocking its proteasomal degradation, causing cleavage of cGAS by caspase-1, resulting in reduced type I IFN signaling and increased ZIKV replication [51]. DENV NS2B3 protease complex cleaves and degrades STING, impairing the production of type I IFN in DENV-infected human monocyte-derived dendritic cells (MDDCs) [52].

Each of the above-mentioned pathways—TLR, RLR, and cGAS—stimulates the transcription and translation of type I IFNs in infected cells, including IFNα, IFNβ, and other subtypes, in an IRF (IRF3 &IRF7) and NF-κB dependent manner, thereby activating and amplifying the antiviral response in both infected and neighboring cells and tissues [50]. Once produced in response to the detection of viral infection, secreted type I IFNs bind to IFNAR1/2, causing heterodimerization of receptor subunits and activating the Janus kinase 1(JAK1) STAT signaling cascade. Phosphorylation of STAT1 and STAT2 leads to the formation of STAT1/2 heterodimers, nuclear translocation, and subsequent complexing with IRF9 [50, 53]. This heterotrimeric complex, known as interferon-stimulated gene factor 3 (ISGF3), binds to IFN-stimulated response elements (ISRE) in the proximal promoter regions of over 100 interferon-stimulated genes (ISGs), upregulating their transcription and translation. The effector functions of ISGs include inhibition of viral entry, degradation of viral proteins or genetic material, and suppression of viral egress [50, 53]. ISGs can act directly to inhibit various stages of the ZIKV and DENV virus life cycle. Inoculation of ZIKV into the eyes of interferon-stimulated gene 15 (ISG15) knockout mice resulted in chorioretinal atrophy, increased retinal cell death, and higher ZIKV replication [54].

Among the ISGs that inhibit *Orthoflavivirus* replication, certain tripartite motif (TRIM) proteins have been identified, with antiviral defenses against viruses, and are also important regulators of the IFN response. TRIM proteins directly inhibit *Orthoflavivirus* replication *via* their ubiquitin-ligase activity. TRIM69 was shown to interact with DENV NS3 protein, inducing its ubiquitination on a specific residue (Lys104), leading to NS3 proteasomal degradation both in mice and human cells [55].

### 3.2 Adaptive immunity to *Orthoflavivirus*

*Orthoflavivirus* infection elicits adaptive immune responses through the activation of both antibodies and T cells. However, the genetic similarity among members of this genus can lead to cross-reactivity in humoral and T-cell responses, which can contribute to a protective effect but also lead to pathological consequences [56, 57].

According to the concept of original antigenic sin, the immune response to a secondary heterotypic DENV infection is shaped by memory B and T cells generated during the primary infection. These cross-reactive cells may have suboptimal avidity for the new serotype, leading to ineffective viral control and contributing to immunopathology and severe DEN disease through the excessive production of inflammatory cytokines [56, 57].

### 3.3 The T cell response against *Orthoflavivirus*

$CD4^+$ T cells modulate viral infection through cytokine secretion, activation of innate immune cells, facilitation of high-affinity antibody production, enhancement of $CD8^+$ T cell responses, immune memory promotion, and direct cytotoxic activity against infected cells [58]. $CD4^+$ T cells differentiate into T helper type 1 (Th1) and follicular helper T (Tfh) cells following viral infections and provide help to $CD8^+$ T cells and B cells [59].

Direct Tfh cells help B cells produce neutralizing antibodies derived from epitopes of C and E proteins. An *ex-vivo* analysis showed expansion and activation of Tfh cells after acute DENV infection in the critical stage of illness, and a correlation between the percentage of activated Tfh cells and the percentage of plasmablasts [60]. DENV-specific $CD4^+$ T cells produce Th1-associated cytokines, including IFN-γ, TNF-α, and IL-2, in response to both infection and vaccination [61]. A subset of $CD4^+$ T cells with surface molecules $CD45RA^+CCR7^-$, known as effector memory T cells re-expressing CD45RA (TEMRA), has been implicated in protective immunity against DENV infection. The progressive expansion of $CD4^+$ TEMRA cells with increasing DENV exposure is associated with a cytotoxic phenotype and elevated expression of CX3CR1, a chemokine receptor linked to cytotoxic activity in both $CD4^+$ and $CD8^+$ T cells [62].

Prior DENV infection or vaccination with tetravalent DENV live-attenuated vaccine (TVLAV) induces T cell responses that recognize ZIKV-derived peptides due to sequence similarity between the viruses. DENV immunity accelerates and amplifies the acute-phase ZIKV-reactive T cell response while modulating its dynamics during convalescence, potentially enhancing viral control [63]. Additionally, prior DENV exposure influences response quality, with ZIKV-specific $CD8^+$ T cells selectively upregulating granzyme B and PD1 in DENV-preexposed individuals [63]. $CD8^+$ T cells can control viral infection through direct cytotoxicity and the production of pro-inflammatory cytokines such as IFN-γ and TNF-α [64]. In DENV infection, $CD8^+$ T cell activation helps to elicit cytotoxic damage by producing IFN-γ, perforin, and granzymes, reducing the viral load [65].

### 3.4 The antibody response against *Orthoflavivirus*

The antibody response to *Orthoflavivirus* infection plays a crucial role in controlling viral infection and dissemination, which is primarily mediated by neutralizing antibodies directed against viral surface glycoproteins. The main targets for the induction of a B-cell and antibody response against *Orthoflavivirus* are the E protein, as well as the prM and NS1 proteins [66]. Following viral infection or vaccination, antigen stimulation of a mature naïve B cell through its B-cell receptor (BCR) triggers activation, leading to the generation of long-lived plasma cells (LLPCs), which produce immunoglobulin G (IgG) neutralizing antibodies, as well as antigen-specific memory B cells (MBCs). During primary *Orthoflavivirus* infection, a rapid but transient increase occurs in antibody-secreting plasmablasts. In the convalescent stage, MBCs and LLPCs contribute to long-term humoral immunity. Upon secondary *Orthoflavivirus* infection, MBCs are characterized primarily by their cross-reactivity to other genetically related Orthoflaviviruses [67].

Antibodies produced during *Orthoflavivirus* infection may contribute to protective Fc-mediated roles, such as complement fixation, antibody-dependent cellular cytotoxicity, and opsonization. Anti-E protein antibodies inhibit viral attachment, internalization, and replication within host cells. The E protein can generate neutralizing antibodies, whereas prM has been implicated in the induction of non-neutralizing antibodies, which can contribute to antibody-dependent enhancement (ADE) and tend to target highly cross-reactive epitopes (this topic will be explored in Section 5.2) [68, 69].

The antigenic epitopes within the E protein are organized into EDI/EDII/EDIII domains. EDIII interacts with attachment factors and receptors, leading to the production of neutralizing antibodies due to its highly antigenic epitopes. Some epitopes within E dimers are conserved among closely related *Orthoflavivirus*, eliciting cross-neutralizing antibodies against viruses from different serogroups [68, 70]. EDI and anti-EDII antibodies are less potent but exhibit broader cross-reactivity across different strains of Orthoflaviviruses compared to EDIII-targeting antibodies [71].

Human monoclonal antibodies against E and prM, isolated from individuals after primary or secondary DENV infection, exhibit either serotype specificity or cross-reactivity. While some potently neutralize DENV infection, others may contribute to ADE [72]. In mice, protective anti-NS1 monoclonal antibodies recognize cell surface-associated forms of NS1, triggering Fc-$\gamma$ receptor-dependent phagocytosis and facilitating the clearance of WNV-infected cells [73].

These findings provide valuable insights for the development of novel vaccines against *Orthoflavivirus* infection, aiming to maximize protective immunity while minimizing the risk of ADE. In this context, the generation of MBCs and the induction of neutralizing antibody responses are essential for controlling viral infection and dissemination, serving as key biomarkers for vaccine efficacy.

## 4. Development of *Orthoflavivirus* vaccines

### 4.1 The YF vaccine as a pioneer

The development of vaccines has been essential for eradicating or controlling diseases that in the past caused millions of deaths annually. YF was considered a serious public health problem, a threat to humanity from the eighteenth century to the early twentieth century, primarily in various tropical regions. There was no specific treatment, nor was there a vaccine to prevent its spread [74]. Therefore, there was a great need for an effective vaccine capable of generating protection against this disease. In 1930, virologist Max Theiler and his team developed the 17D vaccine, obtained from an attenuated strain of the YFV. This was obtained by passing the virus through mouse and chicken embryos, which reduced its virulence without eliminating its ability to induce immunity. This breakthrough was of great importance for the development of effective, safe, and long-term protective vaccines. Theiler received the Nobel Prize in Medicine in 1951 for the development of a vaccine [75]. Currently, the YF vaccine remains one of the most effective, since in most cases a single dose is sufficient to generate lifelong immunity [76].

Once mass vaccination campaigns were implemented for YF in the mid-twentieth century, as was the case in Africa, the great impact that reduced its incidence for decades was evident [77].

### 4.2 *Orthoflaviviridae* vaccines

Despite the global impact of *Orthoflaviviridae* infections, there are still no approved antiviral treatments available. Therefore, it would be extremely important to have a specific treatment for people infected with these viruses, which cause serious health consequences. Given this, it would be essential to develop effective treatments in the future [78]. Vaccines, for their part, have been key to controlling and significantly reducing these infections, which in the past represented a threat to humanity [78, 79].

The most common vaccines against *Orthoflaviviridae* include different types: Attenuated, inactivated, subunit, messenger RNA (mRNA), and viral vector-based.

#### *4.2.1 Attenuated vaccines*

The live-attenuated vaccines (LAVs) against *Orthoflaviviridae* available are for YF, Japanese encephalitis (JE), and DEN, all three of which are mosquito-borne diseases.

With respect to YF, the Asibi strain of this wild-type (WT) virus was isolated from the blood of a Ghanaian man diagnosed with a mild case of YF. Before being used to develop the 17D vaccine, the virus was transferred into rhesus macaques (*Macaca mulatta*) on 53 occasions, with periodic passages in *Aedes aegypti* mosquitoes [80]. For the development of the YF 17D vaccine, it was obtained by 176 serial passages of the WT parental Asibi strain in mouse and chicken embryonic tissues [81]. Currently, neither the original Asibi strain nor the 17D strain is used. The 17D vaccine is obtained from three substrains: 17D-204, 17-213, and 17DD, which are derived from different passages of the original 17D vaccine obtained from the WT Asibi strain (**Figure 3**). However, its safety and efficacy as a YF vaccine have been proven [82].

On the other hand, the SA14-14-2 JE vaccine was licensed in 1989 for vaccination in China and other countries. The WT JE virus strain SA14-14-2 was isolated from a pool of *Culex pipiens* mosquito larvae in Xi'an, China, and the vaccine was derived from this. Briefly, the vaccine was obtained by attenuation of the virus, achieved through periodic passage of the SA14 virus in primary hamster kidney (PHK) cell culture, followed by plaque cloning, resulting in clone 12-1-7. This clone was further attenuated by passage in non-neuronal tissues of mice and Syrian hamsters, as well as by various plaque purifications, resulting in the strain designated SA14-5-3. This was serially transferred subcutaneously into suckling mice, using subcutaneous tissue from the injection site and lymph nodes. Finally, it was cloned in PHK cells to obtain the LAV SA14-14-2 [83]. This licensed vaccine has demonstrated an efficacy of 80% or greater with a single dose and up to 98% with two doses [84].

In Mexico, the first DENV vaccine, developed by Sanofi Pasteur, was licensed in 2015 [85]. The tetravalent vaccine Dengvaxia is based on the use of an attenuated virus consisting of chimeric consisting of premembrane (prM) and envelope (E) structural genes from the four different types of DENV (type 1: Thailand PUO-359/TVP-1140, type 2: Thailand PUO-218, type 3: Thailand PaH881/88, and type 4: Indonesia 1228 [TVP-980]), combined with nonstructural genes from the YF vaccine strain 17D (chimeric DEN YF – CYD) (**Figure 3**) [78, 86]. The Dengvaxia vaccine is licensed in approximately 20 countries, including the United States, several Latin

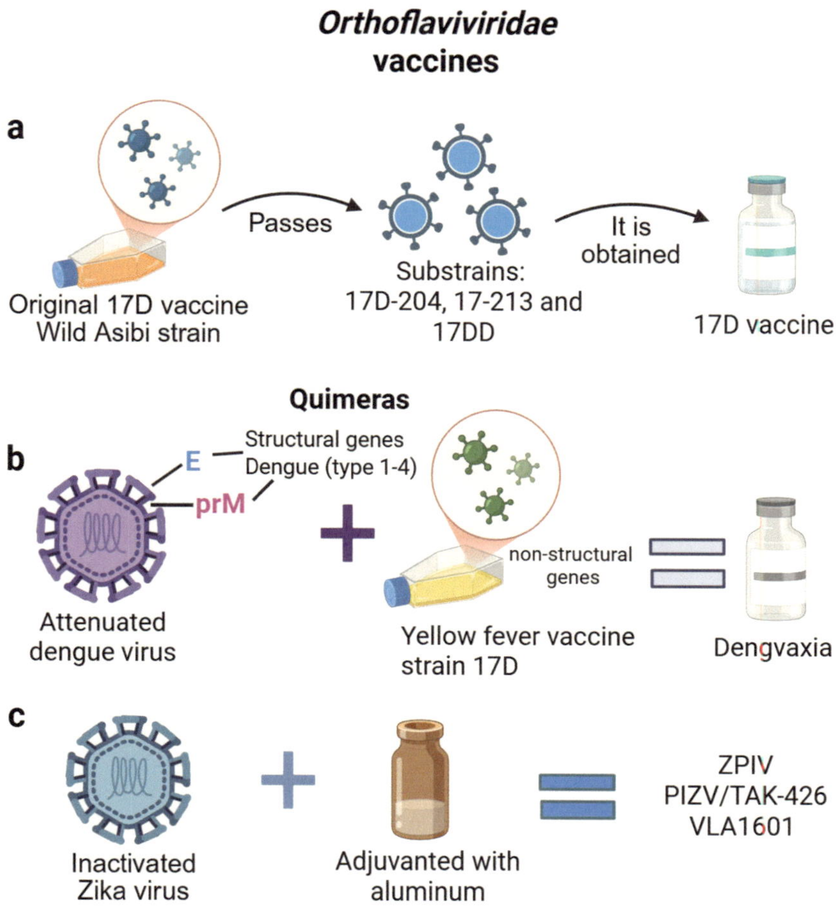


**Figure 3.**
*Process for obtaining different vaccines against the yellow fever virus (YFV), dengue virus (DENV), and Zika virus (ZIKV). a) From the original 17D vaccine obtained from the wild Asibi strain, different passages are carried out and the three substrains are obtained: 17D-204, 17-213, and 17DD, from which the 17D vaccine is obtained. b) The attenuated DENV, which consists of chimeras constituted by structural genes of premembrane (prM) and envelope (E) of the four different types of DENV (types 1–4), combined with nonstructural genes from the yellow fever (YF) vaccine strain 17D (chimeric yellow DEN-CYD), the tetravalent vaccine Dengvaxia is obtained. c) From the inactivated whole ZIKV adjuvanted with aluminum, the vaccines: ZPIV, PIZV/TAK-426, and VLA1601 are obtained.*

American countries, the European Union, Australia, and several Asian nations, solely for the immunization of individuals with prior DENV infection [86]. This is because the vaccine generated controversy due to its variable efficacy and safety risks in individuals not previously infected with the virus [85]. However, due to the lack of a global market for the Dengvaxia vaccine, its manufacturing will be discontinued in 2026, and therefore, its distribution and administration will also cease [86].

In this regard, some examples of FDA-approved attenuated vaccines available against *Orthoflaviviridae* are described in **Table 1**.

### *4.2.2 Inactivated vaccines*

Inactivated vaccines consist of a killed pathogen and are unable to replicate or cause disease [87]. These types of vaccines can induce increased neutralizing antibody titers

| Name of the vaccine | Target virus | Type of vaccine | Manufacturer's name |
|---|---|---|---|
| YFV-17D-204 | YFV | Live-attenuated | Sanofi Pasteur Institute Chiron/ Novartis |
| YFV-17DD | YFV | Live-attenuated | Bio-Manguinhos (Fiocruz) |
| YFV-17D-213 | YFV | Live-attenuated | Federal State Unitary Enterprise of Chumakov Institute |
| SA 14-14-2 | JEV | Live-attenuated | Chengdu Institute of Biological Products |
| IMOJEV/JE-CV | JEV | Live-attenuated | Sanofi Pasteur |
| CYD TVD/ Dengvaxia | DENV | Chimeric live-attenuated | Sanofi Pasteur |
| Qdenga®/TAK 003 (Takeda) | DENV | Live-attenuated | Takeda |
| JE-VAX | JEV | Inactive | The Research Foundation for Microbial Disease of Osaka University |
| IC51/IXIARO | JEV | Inactive | WRAIR |
| EnceVir | Tick-borne encephalitis (TBEV) | Inactive | Microgen |
| TBE-Moscow | TBEV | Inactive | Chumakov Institute of Poliomyelitis and Viral Encephalitides |
| Encepur | TBEV | Inactive | Novartis |
| FSME-IMMUN | TBEV | Inactive | Baxter |

**Table 1.**
*FDA-approved live and inactivated* Orthoflaviviridae *vaccines [4].*

in healthy individuals [88]. However, due to their low immunogenicity, they may largely fail to induce a protective immune response [87, 88]. Therefore, inactivated vaccines often require adjuvants or multiple doses to increase their immunogenicity. However, inactivated vaccines are stable during storage [88] and have a lower risk of vaccine-induced infection, so they are generally considered safer than live vaccines [87].

For the production of inactivated viral vaccines, the pathogen is initially cultured in a substrate for large quantities of antigen production. To obtain substrates for vaccine development, primary cells, fertilized eggs, tissue, and whole organisms were previously used to achieve viral propagation [89, 90]. Currently, researchers are increasingly inclined to use continuous cell lines for virus growth, which has several advantages for these vaccines, such as greater safety, lower production costs, and easier production on a large scale [90]. It should be noted that, after the virus has spread, before inactivation is carried out, it is generally purified and concentrated. To successfully obtain inactivated viruses, physical, chemical, or a combination of both methods is used. There are a large number of inactivation agents or methods, some of which are well-established and others are newly developed. These include: Ethyleneimine derivatives, psoralens, ascorbic acid, hydrogen peroxide, gamma irradiation, heat, UV treatment, among others. However, it should be noted that, for decades, only β-propiolactone (BPL) and formaldehyde have been widely used to obtain authorized inactivated human viral vaccines [91]. It should be noted that vaccine inactivation may cause the epitopes to lose their effect or even cause different subneutralizing antibody reactions [88].

Regarding the inactivated *Orthoflaviviridae* vaccines currently, there are different inactivated whole virus vaccines adjuvanted with aluminum that have been developed against ZIKV, whose commercial name and Developer (in parentheses) are listed below: ZPIV (NIAID/WRAIR/BIDMC), PIZV/TAK-426 (Takeda Pharmaceuticals), VLA1601 (Valneva Austria GmbH), BBV121 (Bharat Biotech International) (**Figure 3**). The mentioned vaccines have completed phase I clinical trials [88].

The FDA-approved inactivated vaccines available against *Orthoflaviviridae* are reported in **Table 1**.

#### *4.2.3 Subunit vaccines*

Subunit vaccines are based on defined antigenic components of microorganisms, administered in conjunction with potent adjuvants. These vaccines contain only the antigenic fragments of the pathogen (a nucleic acid, a polysaccharide, or a protein) that are essential for inducing effective immune responses [92, 93]. If it is a protein, it could be: A purified protein from the pathogen responsible for causing the disease; a synthetic peptide; or a recombinant protein. This is of great importance, as it ensures that the antigen has a well-established composition, is safe for use, and that both its synthesis and purification can be productively expanded [92].

In this context, subunit vaccines against *Orthoflaviviridae*, for example, there are currently more than 30 candidate vaccines based on recombinant antigens that have been developed against DENV. However, although these have shown promising results in preclinical studies, most have not advanced to clinical validation. The only vaccine currently in clinical trials is the tetravalent V180 vaccine (developed by Merck and Co., Kenilworth, NJ, USA). The vaccine is based on truncated versions (DEN-80E) of the DENV1-4 E protein, which is produced in Schneider-2 (S2) cells, obtained from Drosophila, and is used with the ISCOMATRIX™ adjuvant (CSL Behring, King of Prussia, PA, USA) [94].

#### *4.2.4 Messenger RNA (mRNA) vaccines*

Over the past 15 years, there have been major advances in techniques based on the administration of *in vitro*-synthesized mRNA, as well as a significant improvement in its stability, which has impacted the transcendental development of vaccines. The mRNA encoding one or more viral genes can be synthesized *in vitro* and administered into a host organism to transiently promote viral protein expression. The host induces an immune response against the exogenous viral protein and establishes a protective immunological memory [95, 96]. The mRNA is often encapsulated in a lipid nanoparticle (LNP) to protect it from degradation by the host's endogenous nucleases.

Several mRNA-based vaccines, using different approaches, have been developed against different *Orthoflaviviruses*, taking into account multiple antigenic targets. Importantly, these vaccines have been evaluated primarily in preclinical studies against different *Orthoflaviviridae*: DENV, ZIKV, TBEV, and Powassan virus [96]. In the case of ZIKV, mRNA-based vaccines have been developed against this virus, suchv as mRNA-1325 and mRNA-1893, which are in clinical evaluation [97].

#### *4.2.5 Viral vector-based vaccines*

In viral vector-based vaccines, viral vectors are genetically modified to introduce target genes that code for key pathogen antigens. Viral vectors are frequently used today, such as measles virus, adenovirus, measles virus, poxvirus, and influenza virus [98].

Regarding *Orthoflaviviridae* vaccines, only the MV-ZIKV vaccine, developed against the ZIKV, is currently in Phase I clinical trials. It is based on a viral vector from the Schwarz strain of measles that specifically encodes a soluble version of the ZIKV E protein. Although this vaccine has shown high immunogenicity in preclinical studies, information on its immunogenicity and safety at the clinical level is not yet available [4].

### 4.3 Current panorama in the development of vaccines against *Orthoflaviviridae*

The reported efficacy of the various vaccines currently available against *Orthoflaviviridae* varies; however, they generally exhibit a balance with their immunogenicity. Specifically, WNV, DENV, and, in particular, ZIKV have difficulty producing vaccines that induce potent immune responses, which is essential for developing effective vaccines against these viruses.

In recent years, different types of vaccines have achieved significant improvements in immunogenicity and addressed the safety issues faced by various vaccines. However, further progress and research are still needed to address the obstacles to developing certain vaccines [4].

## 5. Immune response and vaccine efficacy

### 5.1 Analysis of immunity induced by *Orthofaviviridae* vaccines: Cellular and humoral responses

The development of vaccines against *Orthoflaviviridae* still faces certain limitations and challenges. Several limitations have hindered the development of vaccines considered safe and effective. This has been particularly linked to the complex immunopathology of certain *Orthoflaviviridae*, in addition to the lack of optimal research tools [99].

For a vaccine to be considered "ideal," it must be safe, even in immunocompromised individuals, and ensure sterile protection against infection. Furthermore, after vaccine administration, it must achieve rapid activation of B cells, which will induce adequate production of protective and neutralizing antibodies. This vaccine must also activate T cells: Helper, memory, and cytotoxic. After vaccination, the CD4+ T cells produced are a key factor in stimulating B cells to produce specific neutralizing antibodies. CD8+ T cells recognize the peptide presented by the major histocompatibility complex (MHC), thereby identifying infected cells for elimination [100, 101]. Long-term protection against various pathogens may be compromised, given the short lifespan of B cells in the blood, and antibody titers decline over time. Therefore, to obtain long-term protection, the immunogen must activate the germinal center of B cells, thereby producing a subpopulation of B cells that, upon re-encounter with the WT or mutated virus, have the capacity to trigger a robust secondary immune response mediated by antibodies. Therefore, to obtain long-term protection, the immunogen must activate the germinal center of B cells, thereby producing a subpopulation of B cells known to trigger a robust antibody-mediated secondary immune response during re-encounter with the WT or mutated virus (**Figure 4**) [102, 103].

Regarding the different *Orthoflaviviridae* vaccines, they trigger different immune responses, so below we will explore some of the most representative examples of vaccines that have been developed over time.

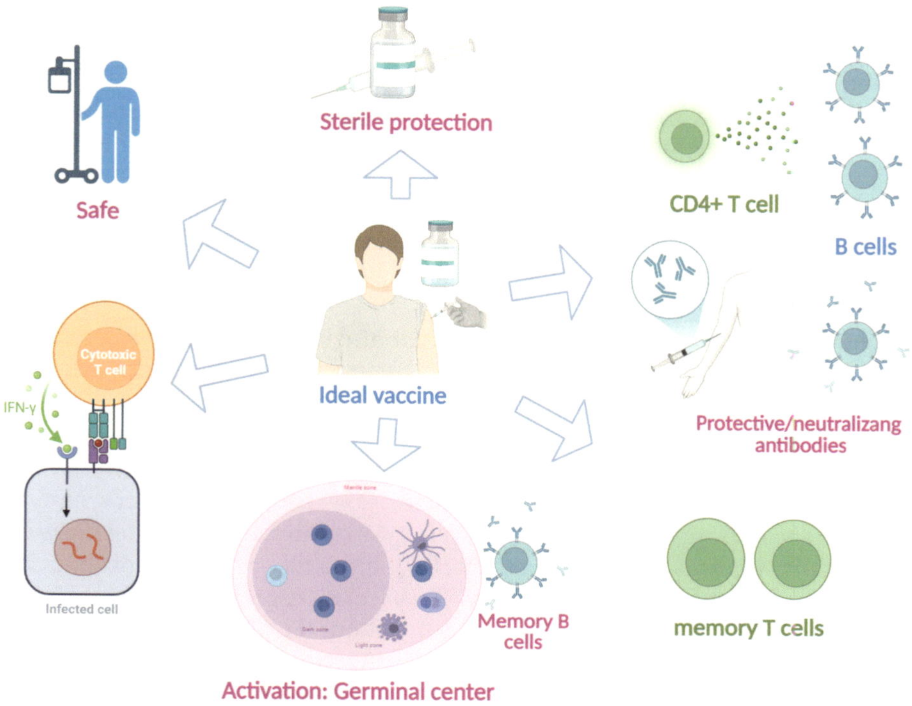


**Figure 4.**
*Characteristics of an ideal vaccine. An ideal vaccine should have the following characteristics: Safety (even in immunocompromised individuals); sterile protection; CD4+ T cell activation for rapid activation and stimulation of B cells to produce protective and neutralizing antibodies against the pathogen; CD8+ T cell activation to recognize and eliminate infected cells through the major histocompatibility complex (MHC); memory T cell activation to sustain future immune responses; and germinal center activation for the generation of memory B cell subpopulations capable of inducing a robust secondary immune response upon re-encounter with the virus (wild or mutated).*

For example, the live-attenuated YF-17D strain vaccine against the YFV gave rise to the development of the 17D-204 and 17DD vaccine substrains, which have been administered to millions of people globally. YF vaccination causes a systemic viral infection that induces the production of lifelong neutralizing antibodies. It also induces a robust T-cell response, similar to acute infections, compared with most other vaccines [104].

Specifically, the overall immune response to the YF17D vaccine is primarily characterized by the stimulation of various defense mechanisms of the innate immune system, which invariably leads to the production of high titers of neutralizing antibodies and a potent activation of both human effector T cells, specifically directed against the YFV, and memory CD8+ T cells. Furthermore, vaccination with YF17D induces TH1 and TH2 lymphocytes that produce a balanced mix of cytokines [105].

On the other hand, in the case of immunization with the IXIARO vaccine against JE, it has been reported that it induces the production of protective neutralizing antibody levels, which can persist up to 60 months after primary vaccination [106].

Regarding the V180 vaccine administered with the ISCOMATRIX™ adjuvant against DENV, six formulations were evaluated in Phase 1 clinical studies, which showed high immunogenicity, compared to formulations administered without aluminum, and two without adjuvant, which showed low immunogenicity. Regarding antibody titers, the geometric mean showed that these generally declined after 6 months, and only for PD3 after 1 year. All nine V180 formulations were generally well tolerated. Notably, V180 formulations, with or without aluminum adjuvant, were associated with fewer adverse events compared to formulations with ISCOMATRIX™ adjuvant, which were associated with a higher number of adverse events [107].

For the ZIKV mRNA-1893 vaccine, in phase 1 clinical studies, this vaccine was reported to be well tolerated at all dose levels evaluated and to have induced robust ZIKV-specific serum neutralizing antibody responses after two doses [97].

Finally, the MV-ZIKA-RSP vaccine candidate, a recombinant attenuated vectored ZIKV vaccine, has been shown to be well tolerated in clinical trials. Immune response is currently under further evaluation [108].

### 5.2 Cross-reactivity between *Orthoflaviviridae* and its relationship with vaccine development

The nature of antigenic variability among *Orthoflaviviruses* and the selection processes responsible for the distribution of variants within populations are defined by the molecular determinants of specificity and cross-reactivity [109]. *Flaviviridae* infection induces the production of a wide variety of antibodies, including those with cross-reactivity to various *Orthoflaviviruses* [110]. It should be noted that in endemic settings where multiple cross-reactive *Orthoflaviviruses* are present, this could interfere with the specific diagnosis of these flavivirus infections. Furthermore, there could also be a risk of ADE of the viral infection, leading to catastrophic clinical outcomes [111].

ADE is a phenomenon in host-pathogen interactions where antibodies, instead of protecting against infection, inadvertently facilitate viral entry into host cells and replication. It is promoted by cross-reactive antibody responses or when sub-neutralizing antibody-virus complexes form, leading to viral escape during phagocytosis by immune cells. This allows the virus to initiate its replication cycle within them, thereby enhancing viral proliferation and its spread to different body tissues and potentially worsening the disease [112, 113].

ADE is associated with worse clinical outcomes in DEN, particularly in cases of secondary infection with a heterotypic DENV strain, where a previously mild and self-limiting infection can progress to a severe form [114]. In contrast, neutralizing antibodies from a primary infection provide protection and long-term immunity in secondary infections caused by a homotypic DENV serotype [115]. Human anti-DENV antibodies cross-react with ZIKV, resulting in both protection through cross-neutralization and ADE due to poorly neutralizing antibodies [116, 117]. **Figure 5** depicts antibody-mediated immunity in DENV infection.

One of the difficulties presented in the development of vaccines directed against *Orthoflaviviridae* is ADE, where previous exposure to an *Orthoflavivirus* can worsen the disease following infection with a different serotype or virus [4]. This phenomenon, combined with the genetic variability among *Orthoflaviviruses*, complicates vaccine design [4, 118]. Ideally, vaccine design and development strategies could leverage these immune responses to promote cross-protection rather than boosting them. Additionally, cross-reactivity of antibodies could be used to strengthen both humoral and cellular immunity, while standardizing immunological assays to define protection thresholds, which is key to promoting future vaccine development [119].

### 5.3 Factors influencing vaccine efficacy

It has been reported that the immune response to vaccination can vary substantially between individuals. The factors involved in influencing cellular and humoral

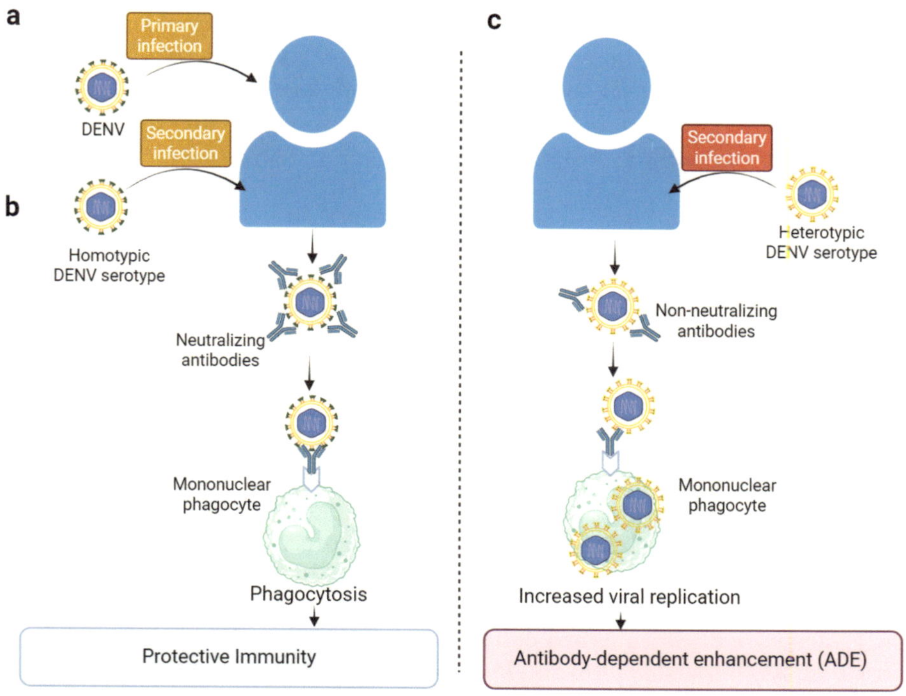


**Figure 5.**
*Antibody-mediated immunity in DENV infection. (a) Primary DENV infection induces neutralizing antibodies at sufficient concentrations to effectively neutralize and eliminate the virus, providing protective immunity. (b) In secondary infection, elicited neutralizing antibodies effectively neutralize DENV infection when it is a homotypic DENV serotype from the primary infection. (c) Antibody-dependent enhancement (ADE) in DENV infection occurs when non-neutralizing antibodies from a primary infection bind to a different DENV serotype during secondary infection, failing to neutralize the virus and facilitating its replication.*

| Factors | Variables involved among individuals |
|---|---|
| Intrinsic of the host | Genetics, sex, comorbidities, pregnancy, age (e.g., perinatal-neonatal (gestational age, birth weight, and "non-experienced immune system") and elderly (immunosenescence and immunocompetence). |
| Extrinsic of the host | Pre-existing immunity, infections, microbiota, and antibiotics. |
| | Perinatal-maternal factors (e.g., placental IgG transfer, maternal immune status, feeding method). |
| | Environmental: Geographic location, season, family size, and exposure to toxins. |
| | Nutritional: Micronutrients, enteropathy, and body mass index. |
| | Behavioral: Exercise, smoking, alcohol consumption, and sleep. |
| About the vaccine | Dose, type of vaccine, adjuvant, and product (specific characteristics of the vaccine used, composition, technological platform, and storage method). |
| Administration | Route (intramuscular (IM), subcutaneous (SC), intradermal (ID), oral (PO), intranasal (IN), etc.), place (anatomical site of vaccine administration, such as the deltoid muscle of the arm or thigh, which may influence absorption and immune response), time of vaccination, schedule, and other co-administered drugs. |

**Table 2.**
*Factors influencing vaccine efficacy.*

responses to vaccines in humans have been extensively studied. These are summarized in **Table 2**. It is important to mention that understanding this wide variety of factors, as well as their impact on both the design and development of vaccine studies and decisions regarding vaccination schedules, offers ways to achieve improved immunogenicity and efficacy of vaccines [120].

## 6. New frontiers in *Orthoflavivirus* immunization

### 6.1 mRNA-LNP vaccines

Several technological advancements have been made in the field of vaccines to improve their efficacy and safety. Nucleic acid-based vaccines utilize DNA or RNA encoding viral antigens, which can be incorporated into a delivery vector and administered to the host [121]. Compared to subunit, inactivated, live-attenuated virus, and DNA-based vaccines, mRNA offers several advantages. mRNA is a noninfectious, non-integrating platform that eliminates the risk of infection and avoids the potential risk of insertional mutagenesis in the host genome [122]. Additionally, mRNA is naturally degraded by cellular processes, and its *in vivo* half-life can be modulated through various modifications and delivery methods [95]. mRNA that encodes viral genes can be synthesized *in vitro* and introduced into a host organism to induce the transient expression of viral proteins. This process triggers an immune response against foreign viral proteins, resulting in the development of protective immunological memory [123].

To enable effective mRNA delivery, several transfection methods have been developed. Naked mRNA is rapidly degraded by extracellular RNases and struggles to enter cells efficiently. Therefore, specialized delivery systems have been designed to enhance mRNA uptake and protect it from degradation [124]. In this context, the application of nanotechnology has revolutionized vaccine development by utilizing nanoparticles, such as lipids, polymers, proteins, and other biomaterials [125]. Nanovaccines combine these nanoparticles with key vaccine components, including antigens and adjuvants, to improve antigen delivery to target cells, leading to broader and more specific protection against several diseases [125]. The mRNA is often encapsulated within LNPs to protect it from degradation by endogenous host nucleases [126].

mRNA vaccines have been developed against *Orthoflaviviruses* using different approaches and antigen targets. Although promising, these mRNA vaccines against DENV, ZIKV, and YFV are currently in the preclinical stage. For instance, mRNA-LNP vaccines have been developed for DENV-1 and DENV-2 using prM and E proteins and have induced high levels of neutralizing antibodies and T-cell immune responses in mice [126, 127]. A multi-target mRNA-LNP vaccine encoding the EIII protein of DENV-1 and DENV-4, as well as the NS1 protein of DENV-2 and DENV-3, has been shown to induce antiviral immune responses and neutralizing antibodies against all four DENV serotypes *in vitro*, without significant ADE [128]. A single low-dose intradermal immunization with an mRNA-LNP vaccine encoding the prM and E proteins of a 2013 ZIKV outbreak strain elicited rapid and durable neutralizing antibody responses in both mice and nonhuman primates [129]. Similarly, an mRNA-LNP for the YFV vaccine encoding the prM, E, and NS1 proteins induced robust innate immune activation and elicited both humoral and cellular immune responses, comparable to the licensed YF-17D vaccine. These responses persisted for up to 6 months in A129 mice and rhesus macaques [130].

### 6.2 Viral vector vaccines

Viral vector vaccines use genetically modified viruses to deliver genes encoding target antigens. These vectors are considered safe and effectively stimulate both innate and adaptive immunity without exposing the host to the complete pathogen [131]. Various viruses from different families, such as *Adenoviridae, Retroviridae, Paramyxoviridae, Rhabdoviridae,* and *Parvoviridae,* have been explored as platforms for viral vector vaccines [132]. These vectors differ in structural characteristics, design strategies, antigen presentation capabilities, immunogenicity, and protective efficacy [131]. Two recombinant adenoviral vector vaccines, cAdVaxD12 and cAdVaxD34, were designed to express the prM and E proteins of all four DENV serotypes. Both induced humoral and cellular immune responses against DENV1–4 [133]. In another study, vaccination with measles virus vector expressing soluble ZIKV E protein significantly reduced infection and prevented fetal loss or damage in an allogeneic mouse pregnancy model [134].

### 6.3 Challenges related to the manufacture and distribution of vaccines in endemic regions

Vaccination is widely regarded as one of the most impactful public health interventions, alongside the provision of clean water and sanitation, in reducing mortality and improving global health [135]. Despite their proven efficacy and cost-effectiveness, numerous challenges hinder the implementation of vaccination programs in low-income countries, limiting their reach and the realization of their full public health potential [136]. *Orthoflavivirus* infections predominantly affect low- and lower-middle-income countries, where inadequate healthcare infrastructure, a shortage of trained medical personnel, and poorly equipped hospitals exacerbate the challenge, creating an urgent demand for affordable and accessible vaccines [137, 138].

Besides the complex immunopathology of *Orthoflavivirus*, one of the major challenges in developing efficient and safe vaccines lies in the time, costs of production, as well as issues related to distribution and availability. Conventional vaccine development typically takes between 8 and 14 years and costs approximately 0.55–1 billion USD, underscoring the urgent need for innovative approaches that enable faster development and regulatory approval to prevent *Orthoflavivirus* infections [139]. In many endemic regions, maintaining the cold chain required for vaccine storage and transportation remains a significant barrier, particularly in remote or rural areas with unreliable electricity supply and limited transport infrastructure, which often results in reduced immunization coverage [140]. Many nations rely heavily on international suppliers, leading to delays in access and vulnerability during outbreaks. This dependence on global supply chains hampers timely responses and restricts the autonomy of affected countries to manage epidemics efficiently [141]. Therefore, the desirable characteristics of a vaccine should go beyond safety and efficacy, encompassing affordability and global accessibility.

### 6.4 Impact of vaccination campaigns against diseases: YF, JE, and DEN

In the early nineteenth century, YF was considered the most dangerous infectious disease, with a high mortality rate. Large-scale vaccination campaigns and widespread elimination of the YF mosquito significantly reduced both YF cases and outbreaks in subtropical and tropical forested regions of Africa and South America for several decades. However, in endemic areas, following a period of low vaccination coverage, YF re-emerged [77, 142]. In this regard, in order to slow the increase in the burden of the disease and control future outbreaks, since 2006 there has been considerable funding for mass preventive vaccination campaigns in the most affected African countries [77]. Currently, contemporary estimates of the burden of YF are lacking. However, a study by Tini Garske [77] determined the impact of recent mass vaccination campaigns, estimating that these campaigns decreased the number of cases and deaths by 27% (95% CI: 22–31%) across the endemic region, achieving reductions of up to 82% in countries where they were implemented.

In another study, national health authorities in Japanese encephalitis-endemic countries were asked for data on JE cases and their vaccination coverage rates. Additionally, data from publications and meetings were included. The results showed that in five countries that introduced the JE vaccine before 2006, the incidence of JE decreased 5–10 years after its introduction, across all ages, ranging from 73 to 100%. Between 2015 and 2021, in six of the countries that introduced the JE vaccine since 2006, the incidence of JE in children under 15 years of age decreased by 14–79%. In Vietnam and Thailand, vaccination programs were reported to have reduced the incidence of acute encephalitis by 74% and 80%, respectively. It should be noted that even the most widely impacted programs took several years to achieve results [143].

In various studies, the CYD-TDV (Dengvaxia®) and Takeda (TAK-003) vaccines (WHO prequalified in 2024) have demonstrated variable long-term efficacy against all four serotypes (DENV-1 to DENV-4), showing robust protection against severe DEN and a promising immune response. However, safety concerns remain, warranting further evaluation [144].

In the specific case of the DENV vaccine Dengvaxia (CYD-TDV) (**Table 1**), which was initially authorized by several endemic countries, this vaccine is intended for people aged 9–45 or 65 years [144]. However, it has only been shown to be effective in people with pre-existing immunity to DENV. People at higher risk of severe disease following vaccination are seronegative individuals, which is why it is contraindicated in people without previous exposure [145, 146].

In a model developed by Ferguson et al. [147], which considered routine vaccination of people aged 2–18 years with 80% coverage, a significant 20–30% reduction in both symptomatic disease and hospitalization was shown in high-transmission populations. However, as mentioned above, due to the lack of a global market for the Dengvaxia vaccine, its manufacturing will be discontinued in 2026, and therefore, its distribution and application will also cease [86].

Therefore, the development of an affordable, effective, and safe DEN vaccine against all four DENV serotypes would represent a major advance in disease control, contributing significantly to achieving the WHO target of reducing DEN morbidity by at least 25% and mortality by at least 50% [144].

## 7. Emerging strategies and approaches to achieve the development of more effective vaccines against *Orthoflaviviridae*

### 7.1 Challenges to creating effective vaccines

The various *Orthoflaviviruses* that cause YE, JE, DEN, and ZIK, among others, have the potential to cause epidemic infections in humans. Therefore, their prevention is essential. Although effective vaccines against YFV and JEV are available, studies to improve preventive measures are still needed [148]. Furthermore, in the current development of vaccines against *Orthoflavivirus*, it would be important to consider the role of nonstructural proteins that could participate as potential targets. Ideally, an important area of study for future vaccine projects is to harness antibody cross-reactivity to enhance vaccine immunogenicity through the induction of humoral and cellular immunity. Likewise, the ability to define host immunity is essential for defining vaccination schedules and thus determining the duration of protection. However, this is limited in non-standardized assays and titer interpretation, which may vary depending on the laboratory. Establishing or predicting a threshold of protection against each key *Orthoflavivirus* is essential for demonstrating vaccination efficacy and determining a basis for promoting vaccination policies before and during epidemics [119].

## 8. Conclusions

*Orthoflavivirus* infections continue to threaten public health, causing significant morbidity and mortality, with DEN being the leading cause. In the absence of effective antiviral treatments, developing safe and efficacious vaccines remains a top priority. Currently, however, only vaccines against YFV and JEV have been approved and demonstrated broad effectiveness. The development of the 17D vaccine for YFV is a notable historical milestone, as it was the first vaccine designed for this genus of viruses and remains one of the most effective. Although a vaccine against DENV (Dengvaxia) is also available and approved in several countries, its use is limited to individuals with prior DENV infection, and its production is set to be discontinued due to insufficient global demand. Various vaccine platforms have been explored over time, including live-attenuated, inactivated, subunit, mRNA-based, and viral vector-based vaccines.

Despite substantial scientific breakthroughs, multiple challenges continue to hinder the development of effective vaccines for other *Orthoflaviviruses* such as DENV, ZIKV, and WNV. These include ADE, antigenic cross-reactivity, and variable immunogenicity, which complicate vaccine design. In addition, the complexity of the immune responses required to ensure both safety and long-lasting protection, as well as the logistical and financial barriers to vaccine production and distribution, particularly in endemic, resource-limited regions, underscore the urgent need for more equitable and innovative approaches. Nevertheless, vaccines remain one of the most powerful and cost-effective tools to prevent infectious diseases. Their strategic use has already demonstrated profound impacts in reducing morbidity and mortality through the use of LAVs, 17D for YF, and SA14-14-2 for JE, and holds promise for other *Orthoflaviviruses* if properly implemented. In addition to the vaccines already approved for clinical use, several experimental candidates have shown great promise in preclinical and early clinical phases. mRNA-based vaccines have yielded encouraging results against ZIKV and DENV in animal models. Additionally, viral vector-based platforms using adenoviruses or measles virus backbones have emerged as viable alternatives, offering strong cellular and humoral responses. While none of these have yet reached clinical approval, they represent a critical pipeline for future vaccine development.

To achieve this goal, a coordinated global effort is essential. Strengthening international collaboration among governments, scientific institutions, industry, and public health organizations will be crucial to accelerate research, facilitate technology transfer, and ensure fair access to vaccines. Investments in local vaccine manufacturing capacity, improvements in cold chain infrastructure, and engagement with communities to enhance vaccine acceptance are equally critical. Only through joint action and sustained commitment will it be possible to effectively control *Orthoflavivirus* infections and prevent future epidemics.

## Acknowledgements

CBF2023-2024-175, IPN-SIP 20250688.

## Conflict of interest

The authors declare no conflict of interest.

## Author details

Samantha Climaco-Arvizu[1] and Mara Gutiérrez-Sánchez[2]*

1 Molecular Oncology Laboratory, Oncology Research Unit, Oncology Hospital, National Medical Center, Instituto Mexicano del Seguro Social (Mexican Institute of Social Security), Mexico City, Mexico

2 Laboratory of Molecular and Cellular Immunobiology, Postgraduate Studies and Research Section, Higher School of Medicine, Instituto Politécnico Nacional (National Polytechnic Institute), Mexico City, Mexico

*Address all correspondence to: mgutierrezsa@ipn.mx

## References

[1] Weaver SC, Barrett ADT. Transmission cycles, host range, evolution and emergence of arboviral disease. Nature Reviews Microbiology. 2004;**2**(10):789-801. DOI: 10.1038/nrmicro1006

[2] Chong HY, Leow CY, Abdul Majeed AB, Leow CH. Flavivirus infection: A review of immunopathogenesis, immunological response, and immunodiagnosis. Virus Research. 2019;**274**:197770. DOI: 10.1016/j.virusres.2019.197770

[3] Pierson TC, Diamond MS. The continued threat of emerging flaviviruses. Nature Microbiology. 2020;**5**(6):796-812. DOI: 10.1038/s41564-020-0714-0

[4] Dutta SK, Langenburg T. A perspective on current flavivirus vaccine development: A brief review. Viruses. 2023;**15**(4):860. DOI: 10.3390/v15040860

[5] Lessa CLS, Hodel KVS, Gonçalves MS, Machado BAS. Dengue as a disease threatening global health: A narrative review focusing on Latin America and Brazil. Tropical Medicine and Infectious Disease. 2023;**8**(5):241. DOI: 10.3390/tropicalmed8050241

[6] World Health Organization. Dengue – Global situation. Geneva: WHO; 2024. Available from: https://www.who.int/emergencies/disease-outbreak-news/item/2024-DON518

[7] Roiz D, Pontifes P, Jourdain F, Diagne C, Leroy B, Vaissière AC, et al. The rising global economic costs of Aedes and Aedes-borne diseases. Science of the Total Environment. 2024;**933**:173054. DOI: 10.21203/rs.3.rs-2679030/v1

[8] Mwaliko C, Nyaruaba R, Zhao L, Atoni E, Karungu S, Mwau M, et al. Zika virus pathogenesis and current therapeutic advances. Pathogens and Global Health. 2021;**115**(1):21-39. DOI: 10.1080/20477724.2020.1845005

[9] World Health Organization. Yellow Fever. Geneva: WHO; 2023. Available from: https://www.who.int/news-room/fact-sheets/detail/yellow-fever

[10] Guzman MG, Halstead SB, Artsob H, Buchy P, Farrar J, Gubler DJ, et al. Dengue: A continuing global threat. Nature Reviews Microbiology. 2010;**8**(Suppl 12):S7-S16. DOI: 10.1038/nrmicro2460

[11] Peinado SA, Aliota MT, Blitvich BJ, Bartholomay LC. Biology and transmission dynamics of Aedes flavivirus. Journal of Medical Entomology. 2022;**59**(2):659-666. DOI: 10.1093/jme/tjab197

[12] Colpitts TM, Conway MJ, Montgomery RR, Fikrig E. West Nile virus: Biology, transmission, and human infection. Clinical Microbiology Reviews. 2012;**25**(4):635-648. DOI: 10.1128/CMR.00045-12

[13] Lu L, Zhang F, Oude Munnink BB, Munger E, Sikkema RS, Pappa S, et al. West Nile virus spread in Europe: Phylogeographic pattern analysis and key drivers. PLoS Pathogens. 2024;**20**(1):e1011880. DOI: 10.1371/journal.ppat.1011880

[14] Clark MB, Schaefer TJ. West Nile virus. In: StatPearls. Treasure Island (FL): StatPearls Publishing; 2023. Available from: https://www.ncbi.nlm.nih.gov/books/NBK555955/

[15] World Health Organization. Dengue: Guidelines for Diagnosis, Treatment,

Prevention and Control. Geneva: WHO; 2009. Available from: https://www.who.int/publications/i/item/9789241547871

[16] Leder K, Torresi J, Brownstein JS, Wilson ME, Keystone JS, Barnett E, et al. Travel-associated illness trends and clusters, 2000-2010. Emerging Infectious Diseases. 2013;**19**(7):1049-1073. DOI: 10.3201/eid1907.121573

[17] Deen JL, Harris E, Wills B, Balmaseda A, Hammond SN, Rocha C, et al. The WHO dengue classification and case definitions: Time for a reassessment. Lancet. 2006;**368**(9530):170-173. DOI: 10.1016/S0140-6736(06)69006-5

[18] Anderson KB, Gibbons RV, Cummings DA, Nisalak A, Green S, Libraty DH, et al. A shorter time interval between first and second dengue infections is associated with protection from clinical illness in a school-based cohort in Thailand. The Journal of Infectious Diseases. 2014;**209**(3):360-368. DOI: 10.1093/infdis/jit436

[19] World Health Organization. Dengue Haemorrhagic Fever: Diagnosis, Treatment, Prevention and Control. 2nd ed. Geneva: WHO; 1997. Available from: http://apps.who.int/iris/bitstream/10665/41988/1/9241545003_eng.pdf

[20] Carod-Artal FJ, Wichmann O, Farrar J, Gascón J. Neurological complications of dengue virus infection. Lancet Neurology. 2013;**12**(9):906-919. DOI: 10.1016/S1474-4422(13)70150-9

[21] Samarasekara K, Munasinghe J. Dengue shock syndrome complicated with acute liver failure and kidney injury, infective endocarditis, and deep vein thrombosis: A case report. Journal of Medical Case Reports. 2018;**12**(1):321. DOI: 10.1186/s13256-018-1862-1

[22] World Health Organization. Zika Virus. Geneva: WHO; 2022. Available from: https://www.who.int/news-room/fact-sheets/detail/zika-virus

[23] Metzler AD, Tang H. Zika virus neuropathogenesis: Research and understanding. Pathogens. 2024;**13**(7):555. DOI: 10.3390/pathogens13070555

[24] World Health Organization. Yellow Fever – Facts and Challenges. Geneva: WHO; 2025. Available from: https://www.who.int/health-topics/yellow-fever/who-yellow-fever-facts-and-challenges#tab=tab_2

[25] Karim SU, Bai F. Introduction to West Nile virus. Methods in Molecular Biology. 2023;**2585**:1-7. DOI: 10.1007/978-1-0716-2760-0_1

[26] Lindenbach B, Rice C. Flaviviridae: The viruses and their replication. In: Knipe DM, Howley PM, editors. Fields' Virology. 4th ed. Philadelphia: Lippincott Williams & Wilkins; 2001. pp. 991-1041

[27] Therkelsen MD, Klose T, Vago F, Jiang W, Rossmann MG, Kuhn RJ. Flaviviruses have imperfect icosahedral symmetry. Proceedings of the National Academy of Sciences of the United States of America. 2018;**115**(45):11608-11612. DOI: 10.1073/pnas.1809304115

[28] Latanova A, Starodubova E, Karpov V. Flaviviridae nonstructural proteins: The role in molecular mechanisms of triggering inflammation. Viruses. 2022;**14**(8):1808. DOI: 10.3390/v14081808

[29] Laureti M, Narayanan D, Rodriguez-Andres J, Fazakerley JK, Kedzierski L. Flavivirus receptors: Diversity, identity, and cell entry. Frontiers in Immunology. 2018;**9**:2180. DOI: 10.3389/fimmu.2018.02180

[30] Mukhopadhyay S, Kuhn RJ, Rossmann MG. A structural perspective of the flavivirus life cycle. Nature Reviews Microbiology. 2005;**3**(1):13-22. DOI: 10.1038/nrmicro1067

[31] Barrows NJ, Campos RK, Liao KC, Prasanth KR, Soto-Acosta R, Yeh SC, et al. Biochemistry and molecular biology of flaviviruses. Chemical Reviews. 2018;**118**(8):4448-4482. DOI: 10.1021/acs.chemrev.7b00719

[32] Ci Y, Shi L. Compartmentalized replication organelle of flavivirus at the ER and the factors involved. Cellular and Molecular Life Sciences. 2021;**78**(11):4939-4954. DOI: 10.1007/s00018-021-03834-6

[33] Seth RB, Sun L, Chen ZJ. Antiviral innate immunity pathways. Cell Research. 2006;**16**(2):141-147. DOI: 10.1038/sj.cr.7310019

[34] McNab F, Mayer-Barber K, Sher A, Wack A, O'Garra A. Type I interferons in infectious disease. Nature Reviews Immunology. 2015;**15**(2):87-103. DOI: 10.1038/nri3787

[35] van Leur SW, Heunis T, Munnur D, Sanyal S. Pathogenesis and virulence of flavivirus infections. Virulence. 2021;**12**(1):2814-2838. DOI: 10.1080/21505594.2021.1996059

[36] Kawai T, Akira S. The role of pattern-recognition receptors in innate immunity: Update on toll-like receptors. Nature Immunology. 2010;**11**(5):373-384. DOI: 10.1038/ni.1863

[37] Wang JP, Liu P, Latz E, Golenbock DT, Finberg RW, Libraty DH. Flavivirus activation of plasmacytoid dendritic cells delineates key elements of TLR7 signaling beyond endosomal recognition. Journal of Immunology. 2006;**177**(10):7114-7121. DOI: 10.4049/jimmunol.177.10.7114

[38] Daffis S, Samuel MA, Suthar MS, Gale M Jr, Diamond MS. Toll-like receptor 3 has a protective role against West Nile virus infection. Journal of Virology. 2008;**82**(21):10349-10358. DOI: 10.1128/JVI.00935-08

[39] Tsai YT, Chang SY, Lee CN, Kao CL. Human TLR3 recognizes dengue virus and modulates viral replication in vitro. Cellular Microbiology. 2009;**11**(4):604-615. DOI: 10.1111/j.1462-5822.2008.01277.x

[40] Suthar MS, Diamond MS, Gale M Jr. West Nile virus infection and immunity. Nature Reviews Microbiology. 2013;**11**(2):115-128. DOI: 10.1038/nrmicro2950

[41] Guo HY, Zhang XC, Jia RY. Toll-like receptors and RIG-I-like receptors play important roles in resisting flavivirus. Journal of Immunology Research. 2018;**2018**:6106582. DOI: 10.1155/2018/6106582

[42] Liang Z, Wu S, Li Y, He L, Wu M, Jiang L, et al. Activation of toll-like receptor 3 impairs the dengue virus serotype 2 replication through induction of IFN-β in cultured hepatoma cells. PLoS One. 2011;**6**(8):e23346. DOI: 10.1371/journal.pone.0023346

[43] Plociennikowska A, Frankish J, Moraes T, Del Prete D, Kahnt F, Acuna C, et al. TLR3 activation by Zika virus stimulates inflammatory cytokine production which dampens the antiviral response induced by RIG-I-like receptors. Journal of Virology. 2021;**95**(10):e01050-20. DOI: 10.1128/JVI.01050-20

[44] Crook KR, Miller-Kittrell M, Morrison CR, Scholle F. Modulation of innate immune signaling by the secreted form of the West Nile virus NS1 glycoprotein. Virology.

2014;**458-459**:172-182. DOI: 10.1016/j.virol.2014.04.036

[45] Town T, Bai F, Wang T, Kaplan AT, Qian F, Montgomery RR, et al. Toll-like receptor 7 mitigates lethal West Nile encephalitis via interleukin 23-dependent immune cell infiltration and homing. Immunity. 2009;**30**(2):242-253. DOI: 10.1016/j.immuni.2008.11.012

[46] Loo YM, Gale M Jr. Immune signaling by RIG-I-like receptors. Immunity. 2011;**34**(5):680-692. DOI: 10.1016/j.immuni.2011.05.003

[47] Fredericksen BL, Keller BC, Fornek J, Katze MG, Gale M Jr. Establishment and maintenance of the innate antiviral response to West Nile virus involves both RIG-I and MDA5 signaling through IPS-1. Journal of Virology. 2008;**82**(2):609-616. DOI: 10.1128/JVI.01305-07

[48] Sui L, Zhao Y, Wang W, Chi H, Tian T, Wu P, et al. Flavivirus prM interacts with MDA5 and MAVS to inhibit RLR antiviral signaling. Cell & Bioscience. 2023;**13**(1):9. DOI: 10.1186/s13578-023-00957-0

[49] Wies E, Wang MK, Maharaj NP, Chen K, Zhou S, Finberg RW, et al. Dephosphorylation of the RNA sensors RIG-I and MDA5 by the phosphatase PP1 is essential for innate immune signaling. Immunity. 2013;**38**(3):437-449. DOI: 10.1016/j.immuni.2012.11.018

[50] Aarreberg LD, Esser-Nobis K, Driscoll C, Shuvarikov A, Roby JA, Gale M Jr. Interleukin-1β induces mtDNA release to activate innate immune signaling via cGAS-STING. Molecular Cell. 2019;**74**(4):801-815.e6. DOI: 10.1016/j.molcel.2019.02.038

[51] Zheng Y, Liu Q, Wu Y, Ma L, Zhang Z, Liu T, et al. Zika virus elicits inflammation to evade antiviral response by cleaving cGAS via NS1-caspase-1 axis. The EMBO Journal. 2018;**37**(18):e99347. DOI: 10.15252/embj.201899347

[52] Aguirre S, Maestre AM, Pagni S, Patel JR, Savage T, Gutman D, et al. DENV inhibits type I IFN production in infected cells by cleaving human STING. PLoS Pathogens. 2012;**8**(10):e1002934. DOI: 10.1371/journal.ppat.1002934

[53] Coldbeck-Shackley RC, Eyre NS, Beard MR. The molecular interactions of ZIKV and DENV with the type-I IFN response. Vaccines (Basel). 2020;**8**(3):530. DOI: 10.3390/vaccines8030530

[54] Singh PK, Guest JM, Kanwar M, Boss J, Gao N, Juzych MS, et al. Zika virus infects cells lining the blood-retinal barrier and causes chorioretinal atrophy in mouse eyes. JCI Insight. 2017;**2**(4):e92340. DOI: 10.1172/jci.insight.92340

[55] Cannac M, Nisole S. TRIMming down flavivirus infections. Viruses. 2024;**16**(8):1262. DOI: 10.3390/v16081262

[56] Rothman AL. Immunity to dengue virus: A tale of original antigenic sin and tropical cytokine storms. Nature Reviews Immunology. 2011;**11**(8):532-543. DOI: 10.1038/nri3014

[57] Vatti A, Monsalve DM, Pacheco Y, Chang C, Anaya JM, Gershwin ME. Original antigenic sin: A comprehensive review. Journal of Autoimmunity. 2017;**83**:12-21. DOI: 10.1016/j.jaut.2017.04.008

[58] Swain SL, McKinstry KK, Strutt TM. Expanding roles for $CD4^+$ T cells in immunity to viruses. Nature Reviews Immunology. 2012;**12**(2):136-148. DOI: 10.1038/nri3152

[59] Maloy KJ, Burkhart C, Junt TM, Odermatt B, Oxenius A, Piali L, et al. CD4(+) T cell subsets during virus infection: Protective capacity depends on effector cytokine secretion and on migratory capability. The Journal of Experimental Medicine. 2000;**191**(12):2159-2170. DOI: 10.1084/jem.191.12.2159

[60] Haltaufderhyde K, Srikiatkhachorn A, Green S, Macareo L, Park S, Kalayanarooj S, et al. Activation of peripheral T follicular helper cells during acute dengue virus infection. The Journal of Infectious Diseases. 2018;**218**(10):1675-1685. DOI: 10.1093/infdis/jiy360

[61] Tian Y, Grifoni A, Sette A, Weiskopf D. Human T cell response to dengue virus infection. Frontiers in Immunology. 2019;**10**:2125. DOI: 10.3389/fimmu.2019.02125

[62] Larbi A, Fulop T. From "truly naïve" to "exhausted senescent" T cells: When markers predict functionality. Cytometry Part A. 2014;**85**(1):25-35. DOI: 10.1002/cyto.a.22351

[63] Grifoni A, Pham J, Sidney J, O'Rourke PH, Paul S, Peters B, et al. Prior dengue virus exposure shapes T cell immunity to Zika virus in humans. Journal of Virology. 2017;**91**(24):e01469-e01417. DOI: 10.1128/JVI.01469-17

[64] Hashimoto M, Im SJ, Araki K, Ahmed R. Cytokine-mediated regulation of CD8 T-cell responses during acute and chronic viral infection. Cold Spring Harbor Perspectives in Biology. 2019;**11**(1):a028464. DOI: 10.1101/cshperspect.a028464

[65] Mapalagamage M, Weiskopf D, Sette A, De Silva AD. Current understanding of the role of T cells in chikungunya, dengue and zika infections. Viruses. 2022;**14**(2):242. DOI: 10.3390/v14020242

[66] Cedillo-Barrón L, García-Cordero J, Bustos-Arriaga J, León-Juárez M, Gutiérrez-Castañeda B. Antibody response to dengue virus. Microbes and Infection. 2014;**16**(9):711-720. DOI: 10.1016/j.micinf.2014.07.011

[67] Adam A, Cuellar S, Wang T. Memory B cell and antibody responses to flavivirus infection and vaccination. Faculty Reviews. 2021;**10**:5. DOI: 10.12703/r/10-5

[68] Slon Campos JL, Mongkolsapaya J, Screaton GR. The immune response against flaviviruses. Nature Immunology. 2018;**19**(11):1189-1198. DOI: 10.1038/s41590-018-0210-3

[69] Hurtado-Monzón AM, Cordero-Rivera CD, Farfan-Morales CN, Osuna-Ramos JF, De Jesús-González LA, Reyes-Ruiz JM, et al. The role of anti-flavivirus humoral immune response in protection and pathogenesis. Reviews in Medical Virology. 2020;**30**(4):e2100. DOI: 10.1002/rmv.2100

[70] Roehrig JT. Antigenic structure of flavivirus proteins. Advances in Virus Research. 2003;**59**:141-175. DOI: 10.1016/s0065-3527(03)59005-4

[71] Chan KR, Ismail AA, Thergarajan G, Raju CS, Yam HC, Rishya M, et al. Serological cross-reactivity among common flaviviruses. Frontiers in Cellular and Infection Microbiology. 2022;**12**:975398. DOI: 10.3389/fcimb.2022.975398

[72] Beltramello M, Williams KL, Simmons CP, Macagno A, Simonelli L, Quyen NT, et al. The human immune response to dengue virus is dominated by highly cross-reactive antibodies endowed

with neutralizing and enhancing activity. Cell Host & Microbe. 2010;**8**(3):271-283. DOI: 10.1016/j.chom.2010.08.007

[73] Chung KM, Thompson BS, Fremont DH, Diamond MS. Antibody recognition of cell surface-associated NS1 triggers Fc-gamma receptor-mediated phagocytosis and clearance of West Nile virus-infected cells. Journal of Virology. 2007;**81**(17):9551-9555. DOI: 10.1128/JVI.00879-07

[74] Monath TP, Vasconcelos PFC. Yellow fever. Journal of Clinical Virology. 2015;**64**:160-173. DOI: 10.1016/j.jcv.2014.11.020

[75] Frierson JG. The yellow fever vaccine: A history. The Yale Journal of Biology and Medicine. 2010;**83**(2):77-85

[76] Beck AT, Barrett ADT. Current status and future prospects of yellow fever vaccines. Expert Review of Vaccines. 2015;**14**(11):1479-1492. DOI: 10.1586/14760584.201510.1586/14760584.2015.1058160

[77] Garske T, Van Kerkhove MD, Yactayo S, Ronveaux O, Lewis RF, Staples JE, et al. Yellow fever in Africa: Estimating the burden of disease and impact of mass vaccination from outbreak and serological data. PLoS Medicine. 2014;**11**(5):e1001638. DOI: 10.1371/journal.pmed.1001638

[78] Mantlo EK, Huang YJ. Knowledge-based mitigation of the environmental risk of Orthoflavivirus live-attenuated vaccines by targeting viral encoded determinants for the mosquito attenuated phenotype. Vaccine. 2025;**54**:127114. DOI: 10.1016/j.vaccine.2025.127114

[79] Cavina L, Bouma MJ, Gironés D, Feiters MC. Orthoflaviviral inhibitors in clinical trials, preclinical in vivo efficacy targeting NS2B-NS3 and cellular antiviral activity via competitive protease inhibition. Molecules. 2024;**29**(17):4047. DOI: 10.3390/molecules29174047

[80] Hahn CS, Dalrymple JM, Strauss JH, Rice CM. Comparison of the virulent Asibi strain of yellow fever virus with the 17D vaccine strain derived from it. Proceedings of the National Academy of Sciences of the United States of America. 1987;**84**:2019-2023. DOI: 10.1073/pnas.84.7.2019

[81] Theiler M, Smith HH. The use of yellow fever virus modified by in vitro cultivation for human immunization. The Journal of Experimental Medicine. 1937;**65**(6):787-800. DOI: 10.1084/jem.65.6.787

[82] Davis EH, Thompson JK, Widen SG, Barrett ADT. Genome characterization of yellow fever virus wild-type strain Asibi, parent to live-attenuated 17D vaccine, from three different sources. Viruses. 2021;**13**(7):1383. DOI: 10.3390/v13071383

[83] Yu Y. Phenotypic and genotypic characteristics of Japanese encephalitis attenuated live vaccine virus SA14-14-2 and their stabilities. Vaccine. 2010;**28**(21):3635-3641. DOI: 10.1016/j.vaccine.2010.02.105

[84] Bista MB, Banerjee MK, Shin SH, Tandan JB, Kim MH, Sohn YM, et al. Efficacy of single-dose SA 14-14-2 vaccine against Japanese encephalitis: A case-control study. Lancet. 2001;**358**(9284):791-795. DOI: 10.1016/s0140-6736(01)05967-0

[85] Harrison LH, L'Azou M. Dengue vaccine: Past, present, and future. Human Vaccines & Immunotherapeutics. 2019;**15**(11):2599-2611. DOI: 10.1080/21645515.2019.1663474

[86] Thomas SJ, Yoon IK. A review of Dengvaxia®: Development to deployment. Human Vaccines & Immunotherapeutics. 2019;**15**(10):2295-2314. DOI: 10.1080/21645515.2019.1658503

[87] Bemben NM, Berg ML. Efficacy of inactivated vaccines in patients treated with immunosuppressive drug therapy. Pharmacotherapy. 2022;**42**(4):334-342. DOI: 10.1002/phar.2671

[88] Peng ZY, Yang S, Lu HZ, Wang LM, Li N, Zhang HT, et al. A review on Zika vaccine development. Pathogens and Disease. 2024;**82**:ftad036. DOI: 10.1093/FEMSPD/FTAD036

[89] Hess RD, Weber F, Watson K, Schmitt S. Regulatory, biosafety and safety challenges for novel cells as substrates for human vaccines. Vaccine. 2012;**30**(17):2715-2727. DOI: 10.1016/j.vaccine.2012.02.053

[90] Barrett PN, Mundt W, Kistner O, Howard MK. Vero cell platform in vaccine production: Moving towards cell culture-based viral vaccines. Expert Review of Vaccines. 2009;**8**(5):607-618. DOI: 10.1586/erv.09.19

[91] Sanders B, Koldijk M, Schuitemaker H. Inactivated viral vaccines. In: Krause PR, editor. Vaccine Analysis: Strategies, Principles, and Control. Berlin: Springer; 2014. pp. 45-80. DOI: 10.1007/978-3-662-45024-6_2

[92] Bill RM. Recombinant protein subunit vaccine synthesis in microbes: A role for yeast? The Journal of Pharmacy and Pharmacology. 2014;**66**(4):566-575. DOI: 10.1111/jphp.12353

[93] Heidary M, Kaviar VH, Shirani M, Ghanavati R, Motahar M, Sholeh M, et al. A comprehensive review of the protein subunit vaccines against COVID-19. Frontiers in Microbiology. 2022;**13**:927306. DOI: 10.3389/fmicb.2022.927306

[94] Araujo SC, Pereira LR, Alves RPS, Andreata-Santos R, Kanno AI, Ferreira LCS, et al. Anti-flavivirus vaccines: Review of the present situation and perspectives of subunit vaccines produced in *Escherichia coli*. Vaccines (Basel). 2020;**8**(3):492. DOI: 10.3390/vaccines8030492

[95] Pardi N, Hogan MJ, Porter FW, Weissman D. mRNA vaccines—A new era in vaccinology. Nature Reviews Drug Discovery. 2018;**17**(4):261-279. DOI: 10.1038/nrd.2017.243

[96] Wollner CJ, Richner JM. mRNA vaccines against flaviviruses. Vaccines (Basel). 2021;**9**(2):148. DOI: 10.3390/vaccines9020148

[97] Essink B, Chu L, Seger W, Barranco E, Le Cam N, Bennett H, et al. The safety and immunogenicity of two Zika virus mRNA vaccine candidates in healthy flavivirus baseline seropositive and seronegative adults: Results of two randomised, placebo-controlled, dose-ranging, phase 1 clinical trials. The Lancet Infectious Diseases. 2023;**23**(5):621-633. DOI: 10.1016/s1473-3099(22)00764-2

[98] Deng S, Liang H, Chen P, Li Y, Li Z, Fan S, et al. Viral vector vaccine development and application during the COVID-19 pandemic. Frontiers in Immunology. 2022;**13**:1007727. DOI: 10.3389/fimmu.2022.1007727

[99] Pollard AJ, Bijker EM. A guide to vaccinology: From basic principles to new developments. Nature Reviews Immunology. 2021;**21**(2):83-100. DOI: 10.1038/s41577-020-00479-7

[100] Rattan A, Richards KA, Knowlden ZAG, Sant AJ. Protein

vaccination directs the CD4+ T cell response toward shared protective epitopes that can be recalled after influenza virus infection. Journal of Virology. 2019;**93**(19):e00947-e00919. DOI: 10.1128/jvi.00947-19

[101] Sharma G, Rive CM, Holt RA. Rapid selection and identification of functional CD8+ T cell epitopes from large peptide-coding libraries. Nature Communications. 2019;**10**:4553. DOI: 10.1038/s41467-019-12493-3

[102] Kräutler NJ, Suan D, Butt D, Bourne K, Hermes JR, Chan TD, et al. Differentiation of germinal center B cells into plasma cells is initiated by high-affinity antigen and completed by Tfh cells. The Journal of Experimental Medicine. 2017;**214**(5):1259-1267. DOI: 10.1084/jem.20161533

[103] Allen CD, Okada T, Cyster JG. Germinal-center organization and cellular dynamics. Immunity. 2007;**27**(2):190-202. DOI: 10.1016/j.immuni.2007.07.009

[104] Bovay A, Fuertes Marraco SA, Speiser DE. Yellow fever virus vaccination: An emblematic model to elucidate robust human immune responses. Current Opinion in Immunology. 2021;**71**:89-96. DOI: 10.1016/j.coi.2021.05.004

[105] Bonaldo MC, Sequeira PC, Galler R. The yellow fever 17D virus as a platform for new live attenuated vaccines. Human Vaccines & Immunotherapeutics. 2014;**10**(5):1256-1265. DOI: 10.4161/hv.28177

[106] Taucher C, Kollaritsch H, Dubischar KL. Persistence of the immune response after vaccination with the Japanese encephalitis vaccine IXIARO® in healthy adults: A five-year follow-up study. Vaccine. 2019;**37**(18):2529-2531. DOI: 10.1016/j.vaccine.2019.03.048

[107] Manoff SB, Sausser M, Russell AF, Martin J, Radley D, Hyatt D, et al. Immunogenicity and safety of an investigational tetravalent recombinant subunit vaccine for dengue: Results of a phase I randomized clinical trial in flavivirus-naïve adults. Human Vaccines & Immunotherapeutics. 2019;**15**(9):2195-2204. DOI: 10.1080/21645515.2018.1546523

[108] Thomas SJ, Barrett A. Zika vaccine pre-clinical and clinical data review with perspectives on the future development. Human Vaccines & Immunotherapeutics. 2020;**16**(10):2524-2536. DOI: 10.1080/21645515.2020.1730657

[109] Frank SA. Specificity and cross-reactivity. In: Frank SA, editor. Immunology and Evolution of Infectious Disease. Princeton (NJ): Princeton University Press; 2002. Chapter 4. Available from: https://www.ncbi.nlm.nih.gov/books/NBK2396/

[110] Vilibic-Cavlek T, Ferenc T, Vujica FM, Bogdanic M, Potocnik-Hunjadi T, Sabadi D, et al. Cross-reactive antibodies in tick-borne encephalitis: Case report and literature review. Antibodies (Basel). 2022;**11**(4):72. DOI: 10.3390/antib11040072

[111] Sauter P, Hober D. Mechanisms and results of the antibody-dependent enhancement of viral infections and role in the pathogenesis of coxsackievirus B-induced diseases. Microbes and Infection. 2009;**11**(4):443-451. DOI: 10.1016/j.micinf.2009.01.005

[112] Uhuami AO, Lawal NU, Bello MB, Imam MU. Flavivirus cross-reactivity: Insights into e-protein conservancy, pre-existing immunity, and co-infection. The Microbe. 2024;**4**:100105. DOI: 10.1016/j.microb.2024.100105

[113] Tirado SM, Yoon KJ. Antibody-dependent enhancement of

virus infection and disease. Viral Immunology. 2003;**16**(1):69-86. DOI: 10.1089/088282403763635465

[114] Morrone SR, Lok SM. Structural perspectives of antibody-dependent enhancement of infection of dengue virus. Current Opinion in Virology. 2019;**36**:1-8. DOI: 10.1016/j.coviro.2019.02.002

[115] Clapham HE, Rodriguez-Barraquer I, Azman AS, Althouse BM, Salje H, Gibbons RV, et al. Dengue virus (DENV) neutralizing antibody kinetics in children after symptomatic primary and postprimary DENV infection. The Journal of Infectious Diseases. 2016;**213**(9):1428-1435. DOI: 10.1093/infdis/jiv759

[116] Dejnirattisai W, Supasa P, Wongwiwat W, Rouvinski A, Barba-Spaeth G, Duangchinda T, et al. Dengue virus sero-cross-reactivity drives antibody-dependent enhancement of infection with zika virus. Nature Immunology. 2016;**17**(9):1102-1108. DOI: 10.1038/ni.3515

[117] Robbiani DF, Bozzacco L, Keeffe JR, Khouri R, Olsen PC, Gazumyan A, et al. Recurrent potent human neutralizing antibodies to Zika virus in Brazil and Mexico. Cell. 2017;**169**(4):597-609.e11. DOI: 10.1016/j.cell.2017.04.024

[118] Ishikawa T, Yamanaka A, Konishi E. A review of successful flavivirus vaccines and the problems with those flaviviruses for which vaccines are not yet available. Vaccine. 2014;**32**(12):1326-1337. DOI: 10.1016/j.vaccine.2014.01.040

[119] Flores HE, Pinzon Burgos EF, Camacho Ortega S, Heredia A, Chua JV. From antibodies to immunity: Assessing correlates of flavivirus protection and cross-reactivity. Vaccines (Basel). 2025;**13**(5):449. DOI: 10.3390/vaccines13050449

[120] Zimmermann P, Curtis N. Factors that influence the immune response to vaccination. Clinical Microbiology Reviews. 2019;**32**(2):e00084-e00018. DOI: 10.1128/cmr.00084-18

[121] Qin F, Xia F, Chen H, Cui B, Feng Y, Zhang P, et al. A guide to nucleic acid vaccines in the prevention and treatment of infectious diseases and cancers: From basic principles to current applications. Frontiers in Cell and Development Biology. 2021;**9**:633776. DOI: 10.3389/fcell.2021.633776

[122] Koppu V, Poloju D, Puvvala B, Madineni K, Balaji S, Sheela CMP, et al. Current perspectives and future prospects of mRNA vaccines against viral diseases: A brief review. International Journal of Molecular and Cellular Medicine. 2022;**11**(3):260-272. DOI: 10.22088/IJMCM.BUMS.11.3.260

[123] Jeeva S, Kim KH, Shin CH, Wang BZ, Kang SM. An update on mRNA-based viral vaccines. Vaccines (Basel). 2021;**9**(9):965. DOI: 10.3390/vaccines9090965

[124] Wadhwa A, Aljabbari A, Lokras A, Foged C, Thakur A. Opportunities and challenges in the delivery of mRNA-based vaccines. Pharmaceutics. 2020;**12**(2):102. DOI: 10.3390/pharmaceutics12020102

[125] Murugan B, Sagadevan S. Nano vaccines: Opportunities and challenges in biomaterial-based vaccine delivery. In: Pradhan M, Yadav K, Singh Chauhan N, editors. Biomaterial-Inspired Nanomedicines for Targeted Therapies. Singapore: Springer; 2024. pp. 101-116. DOI: 10.1007/978-981-97-3925-7_4

[126] Wollner CJ, Richner M, Hassert MA, Pinto AK, Brien JD,

Richner JM. A dengue virus serotype 1 mRNA-LNP vaccine elicits protective immune responses. Journal of Virology. 2021;**95**(12):e02482-20. DOI: 10.1128/JVI.02482-20

[127] Zhang M, Sun J, Li M, Jin X. Modified mRNA-LNP vaccines confer protection against experimental DENV-2 infection in mice. Molecular Therapy - Methods & Clinical Development. 2020;**18**:702-712. DOI: 10.1016/j.omtm.2020.07.013

[128] He L, Sun W, Yang L, Liu W, Li J. A multiple-target mRNA-LNP vaccine induces protective immunity against experimental multi-serotype DENV in mice. Virologica Sinica. 2022;**37**(5):746-757. DOI: 10.1016/j.virs.2022.07.003

[129] Pardi N, Hogan MJ, Pelc RS, Muramatsu H, Andersen H, DeMaso CR, et al. Zika virus protection by a single low-dose nucleoside-modified mRNA vaccination. Nature. 2017;**543**(7644):248-251. DOI: 10.1038/nature21428

[130] Medina-Magües LG, Mühe J, Jasny E, Medina-Magües ES, Roth N, Lopera-Madrid J, et al. Immunogenicity and protective activity of mRNA vaccine candidates against yellow fever virus in animal models. NPJ Vaccines. 2023;**8**(1):31. DOI: 10.1038/s41541-023-00629-7

[131] Wang S, Liang B, Wang W, Li L, Feng N, Zhao Y, et al. Viral vectored vaccines: Design, development, preventive and therapeutic applications in human diseases. Signal Transduction and Targeted Therapy. 2023;**8**(1):149. DOI: 10.1038/s41392-023-01408-5

[132] Henríquez R, Muñoz-Barroso I. Viral vector- and virus-like particle-based vaccines against infectious diseases: A minireview. Heliyon. 2024;**10**(15):e34927. DOI: 10.1016/j.heliyon.2024.e34927

[133] Wu B, Qi Z, Qian X. Recent advancements in mosquito-borne flavivirus vaccine development. Viruses. 2023;**15**(4):813. DOI: 10.3390/v15040813

[134] Nürnberger C, Bodmer BS, Fiedler AH, Gabriel G, Mühlebach MD. A measles virus-based vaccine candidate mediates protection against Zika virus in an allogeneic mouse pregnancy model. Journal of Virology. 2019;**93**(3):e01485-e01418. DOI: 10.1128/JVI.01485-18

[135] Greenwood B. The contribution of vaccination to global health: Past, present and future. Philosophical Transactions of the Royal Society of London. Series B, Biological Sciences. 2014;**369**(1645):20130433. DOI: 10.1098/rstb.2013.0433

[136] MacLennan CA. Vaccines for low-income countries. Seminars in Immunology. 2013;**25**(2):114-123. DOI: 10.1016/j.smim.2013.05.004

[137] Phelan H, Yates V, Lillie E. Challenges in healthcare delivery in low- and middle-income countries. Anaesthesia & Intensive Care Medicine. 2022;**23**(8):501-504. DOI: 10.1016/j.mpaic.2022.05.004

[138] Liang Y, Dai X. The global incidence and trends of three common flavivirus infections (dengue, yellow fever, and Zika) from 2011 to 2021. Frontiers in Microbiology. 2024;**15**:1458166. DOI: 10.3389/fmicb.2024.1458166

[139] Rauch S, Jasny E, Schmidt KE, Petsch B. New vaccine technologies to combat outbreak situations. Frontiers in Immunology. 2018;**9**:1963. DOI: 10.3389/fimmu.2018.01963

[140] Pambudi NA, Sarifudin A, Gandidi IM, Romadhon R. Vaccine cold chain management and cold storage technology to address the challenges of vaccination programs. Energy Reports. 2022;**8**:955-972. DOI: 10.1016/j.egyr.2021.12.039

[141] Pagliusi S, Dennehy M, Homma A. Two decades of vaccine innovations for global public good: Report of the developing countries' vaccine manufacturers network 20th meeting, 21-23 October 2019, Rio de Janeiro, Brazil. Vaccine. 2020;**38**(36):5851-5860. DOI: 10.1016/j.vaccine.2020.05.062

[142] Bifani AM, Ong EZ, de Alwis R. Vaccination and therapeutics: Responding to the changing epidemiology of yellow fever. Current Treatment Options in Infectious Diseases. 2020;**12**:398-409. DOI: 10.1007/s40506-020-00237-2

[143] Letson GW, Marfin AA, Mooney J, Minh HV, Hills SL. JE vaccine global impact assessment team. Impact of vaccination against Japanese encephalitis in endemic countries. PLoS Neglected Tropical Diseases. 2024;**18**(9):e0012390. DOI: 10.1371/journal.pntd.0012390

[144] Okoye EC, Mitra AK, Lomax T, Nunaley C. Dengue fever epidemics and the prospect of vaccines: A systematic review and meta-analysis using clinical trials in children. Diseases. 2024;**12**(2):32. DOI: 10.3390/diseases12020032

[145] Aguiar M, Halstead SB, Stollenwerk N. Consider stopping Dengvaxia administration without immunological screening. Expert Review of Vaccines. 2017;**16**(4):301-302. DOI: 10.1080/14760584.2017.1276831

[146] Aguiar M, Stollenwerk N. Dengvaxia: Age as surrogate for serostatus. The Lancet Infectious Diseases. 2018;**18**(3):245. DOI: 10.1016/s1473-3099(17)30752-1

[147] Ferguson NM, Rodríguez-Barraquer I, Dorigatti I, Mier-Y-Teran-Romero L, Laydon DJ, Cummings DA. Benefits and risks of the Sanofi-Pasteur dengue vaccine: Modeling optimal deployment. Science. 2016;**353**(6303):1033-1036. DOI: 10.1126/science.aaf9590

[148] Hou B, Chen H, Gao N, An J. Cross-reactive immunity among five medically important mosquito-borne flaviviruses related to human diseases. Viruses. 2022;**14**(6):1213. DOI: 10.3390/v14061213

Chapter 5

# Evaluating *Orthoflavivirus* Vaccines with PRNT: Lessons from Yellow Fever Clinical Studies

*Moussa Dia, Ndeye Sakha Bob, Derick Kimathi, Aitana Juan-Giner, George Warimwe, Rebecca F. Grais, Ndongo Dia, Oumar Faye and Gamou Fall*

## Abstract

Flaviviruses represent a major global health concern, with yellow fever virus (YFV) and dengue virus (DENV) among the most studied members of the *Orthoflavivirus* genus. Transmitted by *Aedes* mosquitoes, these viruses are responsible for severe disease outbreaks worldwide. While YFV benefits from an effective vaccine, it remains a threat in endemic regions. Dengue, with its four distinct serotypes, continues to complicate vaccine development and disease management. Plaque assays and the plaque reduction neutralization test (PRNT) are key tools for assessing viral replication and immune responses. PRNT50, commonly used in postvaccination studies, quantifies the serum dilution needed to neutralize 50% of the virus, providing insights into vaccine-induced immunity. In contrast, PRNT90, with a higher neutralization threshold, is more appropriate for diagnostics and epidemiological surveillance, offering a stringent measure of past exposure and protection. This chapter focuses on the application of PRNT in yellow fever (YF) vaccine research, highlighting PRNT50 for evaluating immunogenicity and PRNT90 for population-level immunity assessments. We present findings from a core study validating the PRNT assay, followed by clinical research supporting fractional dosing strategies in various populations, including children and immunocompromised individuals. Together, these studies underscore the value of PRNT in optimizing vaccine use and advancing public health strategies against *Orthoflavivirus* infections.

**Keywords:** *Orthoflavivirus*, yellow fever virus, PRNT50, PRNT90, immunogenicity

## 1. Introduction

Yellow fever virus (YFV), along with dengue virus (DENV), Zika virus (ZIKV), and West Nile virus (WNV), belongs to the *Orthoflavivirus* genus within the *Flaviviridae* family. These arthropod-borne viruses (arboviruses) collectively pose a significant burden on global public health, particularly in tropical and subtropical regions where conditions favor the proliferation of mosquito vectors such as *Aedes aegypti* and *Culex* species. Despite the availability of an effective live-attenuated vaccine (17D) for yellow fever, the disease persists in endemic zones due to factors such as waning immunity, rapid urbanization, ecological changes, and vaccine shortages, all of which contribute to periodic re-emergence and outbreaks [1].

The success of yellow fever control programs hinges on the availability of reliable and standardized tools for detecting confirmed cases through epidemiological surveillance as well as assessing protective immunity in the populations. Neutralizing antibodies serve as the primary correlate of protection against YFV, and their quantification is critical in both clinical and epidemiological contexts [2–4].

The plaque reduction neutralization test (PRNT) remains the gold standard for detecting and quantifying virus-specific neutralizing antibodies. PRNT50, defined as the serum dilution at which a 50% reduction in plaque count is observed, is commonly used in clinical trials to evaluate immunogenicity and to support vaccine licensure [5]. Conversely, PRNT90, indicating a 90% reduction in plaques, offers greater specificity and is often employed in diagnostic and seroepidemiological studies to assess prior exposure or natural immunity in populations [6, 7].

Beyond its role in risk assessment and surveillance, PRNT has become essential in the assessment of dose-sparing strategies aimed at addressing vaccine shortages. The strategy of fractional vaccine dosing using smaller volumes or lower concentrations of antigen has increasingly been adopted during large-scale public health emergencies [8]. Pivotal studies have demonstrated that fractional doses of the 17D YFV vaccine can induce neutralizing antibody responses that are immunologically non-inferior to those elicited by full doses, as measured by PRNT50 [9–11]. PRNT 50 assays have played a critical role in informing WHO recommendations [12] and in generating evidence to address key knowledge gaps, particularly regarding the immunogenicity of various yellow fever vaccines in children and immunocompromised individuals. Notably, the current WHO guidance on fractional dosing excludes children under 2 years of age, pregnant women, and individuals known to be HIV-infected. The results of recent trials, such as the *Immunogenicity and Safety of Fractional Doses of Yellow Fever Vaccines* (YEFE) study and the *Non-Inferiority Fractional Dose Trial for Yellow Fever* (NIFTY), are expected to support a potential revision of these recommendations. Furthermore, tracking geometric mean titers (GMTs) of PRNT responses across multiple time points provides valuable insights into the kinetics, magnitude, and durability of the immune response following vaccination [13].

This chapter provides an in-depth look at the applications of PRNT in yellow fever vaccine research and implementation. We begin with a foundational study that developed and validated the YF-specific PRNT, ensuring consistency and accuracy across laboratories [14]. We then review four key clinical trials that underscore the utility of PRNT in various population groups, including healthy adults, children, and HIV-positive individuals [10, 11, 15, 16].

These studies collectively highlight PRNT's critical role in refining vaccination strategies against *Orthoflavivirus* infections.

## 2. Materials and methods

### 2.1 Development and validation of yellow fever PRNT assay

The PRNT assay described in this section has already been validated and published. The full details of the validation methodology and performance metrics can be found in the original publication [14]. Briefly, PRNT is employed to assess neutralizing antibody responses to YFV. This assay involves several critical steps: virus preparation, serum neutralization, virus-serum incubation, plaque formation, and neutralization titer determination.

#### *2.1.1 Virus and cell preparation*

- *Virus*: The YF 17D vaccine strain was used in this study.
- *Cell line*: Vero cells (ATCC CCL-81) are cultured in Medium 199 with 10% fetal bovine serum (FBS) for virus stock preparation, while porcine stable (PS) kidney cells are cultured in Leibovitz's L-15 medium with 10% fetal bovine serum (FBS) for virus titration and PRNT. Vero cells are maintained in a 37°C, 5% $CO_2$ incubator, and PS cells are used for viral propagation in non-$CO_2$ incubators.
- *Samples*: Serum samples are collected from vaccinated individuals, as well as from human specimens obtained through YFV surveillance efforts. The samples are heat-inactivated at 56°C for 30 minutes to prevent complement activity and non-specific binding.
- *Dilution*: Serum samples are serially diluted from 1:10 to 1/20480 to assess the neutralizing antibody titer across.

#### *2.1.2 Neutralization assay procedure*

1. *Neutralization step*: In a 24-well format, each diluted serum sample is mixed with a fixed amount of YFV (typically 100–200 PFU). The mixture is incubated for 1–2 hours at 37°C to allow the neutralizing antibodies present in the serum to bind and neutralize the virus.
2. *Plaque formation assay*: After incubation, the virus-serum mixture is added to PS cell monolayers. The cells are incubated at 37°C for 3–4 hours to allow virus adsorption.
3. *Overlay and incubation*: The cells are overlaid with carboxymethyl cellulose containing 3% FBS, and the plates are incubated at 37°C for 4–5 days to allow plaques to form. The overlay ensures that viral plaques are isolated and distinct.
4. *Plaque counting*: After the incubation period, the cells are fixed and stained with Amido black solution. Plaque counts are performed visually, and the reduction in plaque formation due to neutralization by the serum is determined.

#### *2.1.3 Performance parameters of the PRNT assay*

##### *2.1.3.1 Validation of the PRNT assay*

To validate the PRNT assay, several performance metrics were assessed:

- *Intra-assay precision*: Reproducibility of the assay within a single run, assessed by performing the assay with 10 repeats of a control serum sample.

- *Inter-assay precision*: Consistency of results across different assay runs, evaluated by testing the same control serum on different days and with different operators.
- *Specificity*: The assay's ability to detect YFV-specific neutralizing antibodies without cross-reactivity with other flaviviruses. This was tested using sera from individuals infected with other flavivirus species (dengue virus, Zika virus, etc.).
- *Lower limit of quantification (LLOQ)*: The sensitivity of the assay in detecting low-level neutralizing antibodies. This was determined by analyzing control sera with known low titers and assessing the assay's ability to accurately detect these low levels.
- *Dilutability*: The ability of the assay to provide consistent results when serum samples are serially diluted. This was evaluated by performing the PRNT on samples at different dilution levels to ensure that neutralization titers remain stable.

#### *2.1.3.2 Control range of acceptability*

A critical element in the validation process was defining the control range of acceptability. To ensure consistent and reliable assay performance, a control serum (with known neutralizing antibody titers) was used as a benchmark for each assay. The GMT of 10 repeated tests was calculated, and the control range of acceptability was defined as GMT ± 3 standard deviations (SD). This range ensures that the assay provides accurate, reproducible results across different runs, with acceptable variation.

#### *2.1.3.3 Application to field and clinical samples*

Once validated, the PRNT assay was tested using a variety of samples:

- *International standard serum*: Used as a benchmark to calibrate the assay's performance.
- *Vaccinee sera*: Sera from individuals who had received the yellow fever vaccine were tested to assess the immune response postvaccination.
- *Clinical samples*: Sera from individuals participating in yellow fever surveillance studies were analyzed to evaluate the assay's effectiveness in real-world diagnostic settings.

To assess the immunogenicity and operational feasibility of fractional dosing of yellow fever vaccines across various populations, four key clinical trials were conducted in diverse settings using the validated PRNT50 assay. Each study addressed a specific demographic group and contributed critical data supporting the broader implementation of dose-sparing strategies.

### 2.2 Multi-vaccine fractional dose study in adults

Following the first large use of YFV vaccines in Kinshasa, the Democratic Republic of the Congo [9], a randomized, double-blind, non-inferiority trial was conducted in Uganda and Kenya to assess the immunogenicity and safety of fractional (one-fifth) doses of four WHO-prequalified yellow fever vaccines in adults aged 18–59 years. In this context, fractional administration refers to the delivery of 0.1 mL, equivalent to one-fifth of the standard 0.5 mL dose, *via* intramuscular injection, in line with WHO emergency recommendations to extend vaccine supply. These four vaccines are all derived from the 17D strain and include 17DD from Bio-Manguinhos/Fiocruz (Brazil); 17D-213 from the Federal State Unitary Enterprise of the Chumakov Institute of Poliomyelitis and Viral Encephalitides (Russia); and two 17D-204 formulations from the Institut Pasteur de Dakar (Senegal) and Sanofi Pasteur (France). The study was sponsored by MSF/Epicenter and implemented in collaboration with the KEMRI-Wellcome Trust Research Programme and the Institut Pasteur de Dakar (IPD). IPD served as the Regional Reference Laboratory, applying the validated PRNT assay (described above) to measure neutralizing antibody titers by PRNT 50 at 28 days postvaccination. Results of this study have been published [10].

#### 2.2.1 Study design

This was a multicenter, randomized, double-blind, non-inferiority trial designed to compare the immunogenicity of fractional doses (one-fifth of the standard dose) of four WHO-prequalified yellow fever vaccines versus the standard dose. The one-fifth fraction (0.1 mL instead of 0.5 mL) was selected based on previous operational experience during emergency vaccination campaigns (notably in Kinshasa, 2016) and on preliminary evidence suggesting that this volume could maintain immunogenicity while significantly extending vaccine supply. It also aligns with WHO recommendations during periods of global vaccine shortage.

The study included adults aged 18–59 years from Uganda and Kenya, recruited based on inclusion criteria such as no prior yellow fever vaccination and no active infection.

#### 2.2.2 Sample size

A total of 960 participants were enrolled and randomly assigned to vaccine manufacturer and dose (120 per group), with 480 participants receiving a standard dose and 480 participants receiving a fractional dose of one of the four vaccines.

#### 2.2.3 Immunogenicity assessment

Immunogenicity was evaluated using PRNT50, with seroconversion defined as a fourfold increase in neutralizing antibody titers compared to baseline. The primary endpoint was seroconversion at 28 days postvaccination, with a non-inferiority margin of −10%, based on the assumption that at least 95% of participants would respond to a standard dose. Secondary outcomes included seroconversion at days 10 and 365, as well as comparisons of geometric mean titers (GMTs) and geometric mean fold increases (GMFIs) between fractional and standard dose groups. Neutralization assays were performed at a reference laboratory using a standardized 17D-204 YFV strain.

#### 2.2.4 Statistical analysis

Analysis was performed using intention-to-treat and per-protocol populations. A non-inferiority criterion was considered if the lower bound of the 95% CI for the difference in seroconversion rates was above −10%.

### 2.3 Pediatric evaluation of fractional doses (17D-213 vaccine)

The second phase of the YEFE trial aimed to evaluate the immunogenicity and safety of fractional doses (one-fifth of the standard dose) of the 17D-213 yellow fever vaccine in children aged 9–59 months. The randomized, double-blind, non-inferiority substudy took place in Uganda. This substudy was implemented following the main outcomes of the trial in adults. The substudy was sponsored by MSF/Epicenter and implemented in the Epicenter Research Centre in Mbarara, Uganda, a country endemic for yellow fever. Laboratory analyses were conducted at the Institut Pasteur de Dakar (IPD), which acted as the Regional Reference Laboratory. The validated PRNT50 assay developed at IPD was used to measure seroconversion rates 28 days postvaccination. The study findings were published [15].

#### *2.3.1 Study design*

A randomized, blinded, non-inferiority trial evaluated fractional doses in children aged 9–59 months. The study aimed to assess the immunogenicity of fractional (one-fifth) doses of the 17D-213 yellow fever vaccine compared to the standard dose. The use of the one-fifth fractional dose was based on prior adult studies and WHO emergency recommendations, which showed that this volume was sufficient to induce protective immune responses while significantly conserving vaccine supplies during global shortages.

#### *2.3.2 Sample size*

A total of 420 children were enrolled and randomly assigned to receive either the standard dose (0.5 mL) or a fractional dose (one-fifth of the standard, i.e., 0.1 mL) of the vaccine.

#### *2.3.3 Immunogenicity assessment*

Seroconversion rates were measured by PRNT 50 at 28 days postvaccination, with additional assessments at 10 days and 1 year to evaluate early and long-term immune responses.

#### *2.3.4 Statistical analysis*

A non-inferiority framework was used to compare seroconversion rates between groups, with the primary endpoint set at 28 days postvaccination. In the pediatric YEFE substudy, it was assumed that at least 90% of children receiving a standard dose would seroconvert by day 28.

### 2.4 Fractional dosing in HIV-positive adults (17D-213 vaccine)

This randomized, double-blind, non-inferiority substudy of the YEFE program assessed the one-fifth fractional dose of the 17D-213 yellow fever vaccine in HIV-positive adults (CD4 $\geq$ 200 cells/μL) in Kenya. The trial was sponsored by MSF/Epicenter and implemented by the KEMRI-Wellcome Trust Research Programme in Kilifi, Kenya. Laboratory testing was conducted at the Institut Pasteur de Dakar, which provided PRNT50 testing as the Regional Reference Laboratory. The validated

PRNT assay confirmed non-inferior immune responses at day 28 and 1 year postvaccination. Findings were already published [16].

### *2.4.1 Study design*

A randomized, double-blind, non-inferiority study focused on HIV-positive adults with CD4+ T-cell counts ≥200 cells/μL. Participants received either the standard or one-fifth of the 17D-213 yellow fever vaccine.

### *2.4.2 Sample size*

A total of 250 HIV-positive participants aged 18–59 years were enrolled.

### *2.4.3 Immunogenicity assessment*

Seroconversion was assessed by PRNT50 at 28 days, 10 days, and 1 year postvaccination to evaluate both short-term and long-term immunity.

### *2.4.4 Statistical analysis*

Non-inferiority was determined by comparing the seroconversion rates of the fractional dose group with the standard dose group at each time point.

## 2.5 Minimum effective dose study (Institut Pasteur de Dakar 17D-204 vaccine)

This randomized, double-blind, non-inferiority trial aimed to identify the lowest effective dose of the 17D-204 yellow fever vaccine produced by the Institut Pasteur de Dakar. Conducted in Kenya and Uganda, the study enrolled healthy adults aged 18–59 years, evaluating fractional doses of 1000 International Units (IU), 500 IU, and 250 IU compared to the standard dose of 13,803 IU. Sponsored by Oxford University and conducted in partnership with KEMRI-Wellcome Trust, Epicenter, the Institut Pasteur de Dakar, and UVRI, the study relied on PRNT 50 testing performed by the Regional Reference Laboratory. The results demonstrated that doses as low as 500 IU maintained non-inferior immunogenicity. The study findings were published [11].

### *2.5.1 Study design*

This study aimed to define the lowest effective dose of the Institut Pasteur de Dakar 17D-204 yellow fever vaccine capable of inducing protective immunity. Participants were randomized into four groups, receiving the standard dose (13,803 IU) or fractional doses (1000 IU, 500 IU, and 250 IU).

### *2.5.2 Sample size*

A total of 480 participants (120 per group) were enrolled in this double-blind, randomized, non-inferiority trial at sites in Kenya and Uganda. Eligible adults aged 18–59 years were either HIV-negative or clinically stable HIV-positive on antiretroviral therapy with CD4 counts above 200 cells/μL. Participants had no prior yellow fever vaccination or infection, were not pregnant or lactating, and agreed to follow trial procedures.

| Study* | Objective | Population and sample size | Vaccine(s)** | Statistical design / Reference |
|---|---|---|---|---|
| YEFE trial | Multi-vaccine Fractional Dose Study in Adults Assess immunogenicity and safety of one-fifth fractional doses | Adults aged 18–59 years (N = 960; 120 per dose group × 4 vaccines; standard vs. one-fifth dose) | 17DD (Fiocruz), 17D-213 (Chumakov), 17D-204 (IPD and Sanofi) | Double-blind, randomized, non-inferiority Juan-Giner et al. [10] |
| YEFE substudy Pediatric (17D-213) | Evaluate the immunogenicity of fractional doses in children | Children aged 9–59 months (N = 420; standard vs. one-fifth dose) | 17D-213 (Chumakov) | Double-blind, randomized, non-inferiority Juan-Giner et al. [15] |
| YEFE substudy HIV-positive Adults (17D-213) | Assess immunogenicity in the HIV+ population | HIV-positive adults aged 18–59 years (N = 480; 120 per dose group; standard vs. one-fifth dose) (stable ART and CD4 > 200) | 17D-213 (Chumakov) | Double-blind, randomized, non-inferiority Kimathi et al. [16] |
| NIFTY Trial Minimum Effective Dose Adults (17D-204 Vaccine) | Determine the minimum effective dose of the YF vaccine | Adults aged 18–45 years (1000 IU, 500 IU, and 250 IU doses vs. standard dose) | 17D-204 (Institut Pasteur de Dakar) | Double-blind, randomized, non-inferiority Kimathi et al. [11] |

*Immunogenicity in all studies was assessed using PRNT50.
**17DD from Bio-Manguinhos-Fiocruz (Brazil), 17D-213 from Federal State Unitary Enterprise of Chumakov Institute of Poliomyelitis and Viral Encephalitides (Russia), and 17D-204 from Institut Pasteur Dakar (Senegal) and Sanofi Pasteur (France).*

**Table 1.**
*Summary of yellow fever vaccine fractional dose trials and sub-studies, showing key differences in target population, vaccine types, endpoints, and analytical approaches.*

### *2.5.3 Immunogenicity assessment*

Seroconversion was measured by PRNT50 at 28 days postvaccination and at 10 days, 1 year, and 2 years postvaccination.

### *2.5.4 Statistical analysis*

Non-inferiority analyses were performed comparing each fractional dose to the standard dose, using the 10% non-inferiority margin.

To facilitate comparison across the different studies discussed above, the main characteristics—including objectives, populations, sample sizes, immunogenicity assays, and statistical approaches—are summarized in **Table 1**.

## 3. Results

### 3.1 Development and validation of yellow fever PRNT assay

The yellow fever PRNT 90 demonstrated excellent performance metrics: 100% intra-assay precision, 95.6% inter-assay precision, and 100% specificity. The LLOQ was found to be reliable, with a dilutability rate of 95.3%, confirming that the assay could provide accurate and reproducible results, suitable for both clinical and surveillance settings [14]. Comparison with PRNT 50 metrics demonstrates some variability in inter-assay precision and specificity but remains very effective for general immunogenicity testing (publication ongoing) (**Figure 1**). For all evaluated performance parameters of the PRNT50 assay, the threshold of 80% was exceeded, confirming the assay's reliability for immunogenicity assessment.

Calibration curves for PRNT 50 and PRNT 90 were generated using GMT values from 10 replicates, with control ranges defined by the mean and corresponding ±3 standard deviations (SD). These curves confirmed assay consistency, as all measured titers remained within the expected thresholds, supporting reliable interpretation of subsequent performance parameters.

The robustness of the PRNT assay was further demonstrated by establishing an acceptability range for the international standard control (NIBSC), calculated from the same dataset. This range served as a reference for routine assay validation and enabled early detection of deviations across test runs (**Figure 2**).

### 3.2 Multi-vaccine fractional dose study in adults

The study demonstrated that fractional doses (one-fifth of the standard dose) of all four WHO-prequalified yellow fever vaccines were non-inferior to full doses at 28 days postvaccination. Using PRNT50 as the primary endpoint for immunogenicity assessment, seroconversion rates in the fractional dose groups were consistently high and comparable to those in the standard dose groups, with all meeting the predefined non-inferiority margin of −10%. These findings support the use of fractional dosing as a dose-sparing strategy during outbreaks, particularly in resource-limited settings.

The detailed findings of this study are reported in full in the original publication by Juan-Giner et al. [10].

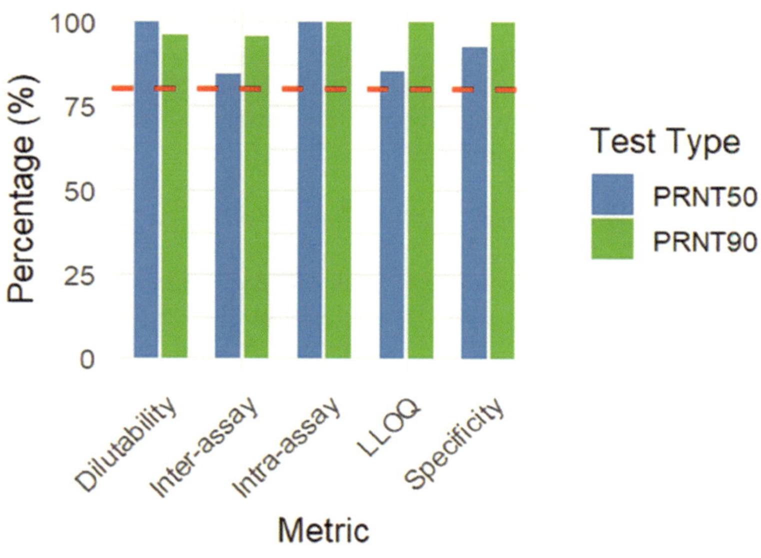


**Figure 1.**
*Performance comparison of PRNT 50 and PRNT 90.*

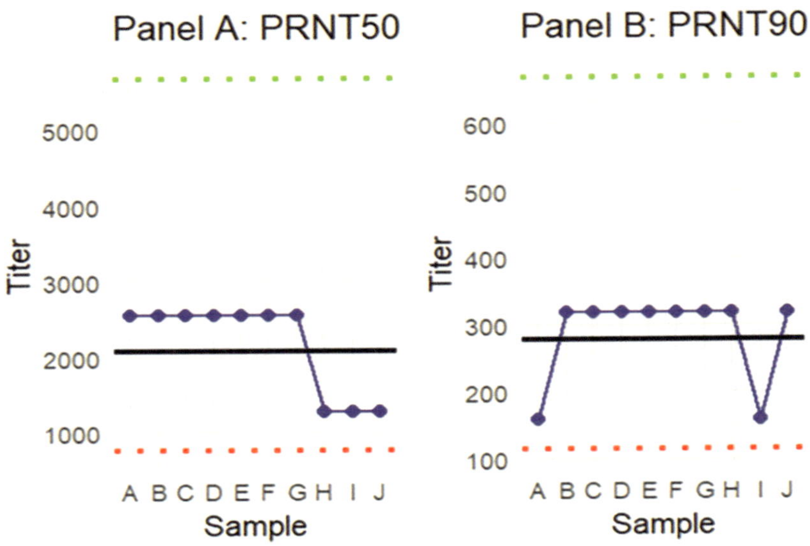


**Figure 2.**
*Control calibration acceptability range for PRNT50 and PRNT90. Calibration curves were generated for both PRNT50 (Panel A) and PRNT90 (Panel B) using titers obtained from 10 independent replicates (ranging from 1280 to 2560 for PRNT50 and 160 to 320 for PRNT90). Geometric mean titers (GMTs) and standard deviations were calculated, defining acceptability ranges as GMT ± 3 SD to ensure quality control across assays.*

### 3.3 Pediatric evaluation of fractional doses (17D-213 vaccine)

PRNT50 seroconversion rates in the fractional dose group were 97% at day 28, slightly lower than the standard dose (100%). However, the fractional dose group achieved non-inferiority, suggesting that reduced vaccine doses are effective for children [15].

### 3.4 Fractional dosing in HIV-positive adults (17D-213 vaccine)

The study showed that fractional doses of 17D-213 were non-inferior to the standard dose in HIV-positive adults. The results at both 28 days and 1 year postvaccination showed sustained immune responses, demonstrating that fractional dosing is a viable option for immunocompromised individuals [16].

### 3.5 Minimum effective dose study (Institut Pasteur de Dakar 17D-204 vaccine)

The study found that doses as low as 500 IU were non-inferior to the standard dose in terms of seroconversion rates. PRNT50 results from all dose groups demonstrated comparable immune responses, highlighting the potential for dose optimization in vaccine campaigns [11].

## 4. Discussion

The increasing frequency of *Orthoflavivirus* outbreaks has highlighted the need for effective vaccination strategies and reliable immunological assessment tools. For instance, the World Health Organization (WHO) reported over 7.6 million dengue cases globally as of April 2024, with the Region of the Americas experiencing an unprecedented surge, surpassing the annual high of 4.6 million cases recorded in 2023. Furthermore, the global incidence of dengue has seen a 30-fold increase over the past 50 years, placing two-fifths of the world's population at risk. These trends underscore the critical need for robust vaccination programs and precise immunological assessments to mitigate the escalating impact of *Orthoflavivirus* infections [17, 18]. The PRNT remains a cornerstone in evaluating the immunogenicity of yellow fever (YF) vaccines, demonstrating consistent reliability across diverse clinical scenarios [14]. Among its formats, PRNT 50 is most commonly used in clinical trials to assess vaccine-induced immune responses, offering a balance between sensitivity and practicality. In contrast, PRNT 90, a more stringent format, is preferred in diagnostic settings to reduce cross-reactivity among closely related flaviviruses and ensure greater specificity in serological surveillance [19].

### 4.1 YF PRNT validation and performance

A pivotal study focused on the validation of the PRNT assay for YF demonstrated its robustness and reproducibility. The assay exhibited high intra- and inter-assay precision, confirming its suitability for both diagnostic purposes and vaccine clinical trials. These findings established PRNT as a reliable method for assessing neutralizing antibody responses in various settings [5–7, 14].

### 4.2 Vaccine fractional dose strategies in diverse populations

In response to vaccine shortages, fractional dosing of the YVF vaccine has been explored as a viable strategy. Clinical trials have shown that fractional doses of YF vaccines elicit immune responses comparable to standard doses. For instance, a study in adults demonstrated that doses as low as 500 International Units (IU) were non-inferior to the standard 13,803 IU dose in achieving seroconversion [11]. Similarly, research involving HIV-positive individuals indicated that one-fifth fractional doses of the 17D-213 YF vaccine were sufficiently immunogenic and safe, meeting non-inferiority criteria compared to standard doses [16].

In pediatric populations, fractional dosing has also proven effective, with studies reporting high seroconversion rates in children receiving one-fifth doses. These findings support the use of fractional dosing in mass vaccination campaigns during vaccine shortages. However, additional data are needed to confirm the long-term durability of protection in this age group [15].

### 4.3 Operational insights and public health implications

The implementation of vaccine fractional doses during mass vaccination campaigns has provided critical operational insights. For example, during a preemptive campaign in the Democratic Republic of Congo, fractional doses of the 17DD YF vaccine were administered, demonstrating effective immunogenicity and informing future outbreak responses. These experiences underscore the practicality of fractional dosing in resource-limited settings and during vaccine shortages [10, 20].

### 4.4 Advancements in automated PRNT for enhanced efficiency and precision

The growing demand for scalable, high-throughput neutralization assays has driven the development of automated PRNT platforms, which enhance both the efficiency and reproducibility of results. These automated systems reduce human error and significantly accelerate analysis, enabling large-scale screening, making them particularly valuable for epidemiological studies and vaccine evaluations. Notable advancements in this field are the incorporation of fluorescent PRNT (fluoPRNT), alternative fluorescence-based serological assay ($SNT^{FLUO}$), and fluorescence reduction neutralization assay (FRNA), which replaces traditional plaque counting with fluorescent markers to detect virus neutralization [21, 22]. This innovative approach not only improves sensitivity and precision but also accelerates the process, especially in high-volume testing scenarios, providing a faster and more reliable alternative to conventional methods. The integration of fluorescent markers into PRNT represents a major leap forward in immunological assays, offering enhanced accuracy and efficiency for virus-neutralizing antibody detection [21–23].

## 5. Conclusion

Research surrounding *Orthoflavivirus*, particularly in the context of yellow fever and related flaviviruses, underscores the critical role of immunological tools like the PRNT in advancing vaccine evaluation and epidemiological surveillance. As outbreaks of *Orthoflavivirus* continue to pose significant global health risks, the refinement of neutralization assays, including PRNT and emerging technologies such as fluoPRNT, is essential to improve the speed, sensitivity, and scalability of diagnostics. By integrating high-throughput methods and automation, we can enhance the accuracy of immune response assessment, supporting more effective vaccine strategies and public health responses. The ongoing development of these tools, alongside continued research into the nuances *of Orthoflavivirus* immunity, is crucial in addressing future challenges and shaping vaccine policy worldwide.

## Acknowledgements

We would like to express our sincere gratitude to all team members at the WHO Collaborating Center for Arboviruses and Hemorrhagic Fever Viruses who contributed to the development, implementation, and analysis of this work. Special thanks go to the PRNT team—Djibril Ndione, Pape Ibrahima Diop, Diarietou Bousso Tall, Abou Diallo, Franck William Mendy, and Mame Daba Thiam—for their technical expertise and dedication to the PRNT assay validation and clinical studies.

## Conflict of interest

No conflict of interest for any of the authors was declared.

## Author details

Moussa Dia[1*], Ndeye Sakha Bob[1], Derick Kimathi[2], Aitana Juan-Giner[3], George Warimwe[2], Rebecca F. Grais[3], Ndongo Dia[1], Oumar Faye[1] and Gamou Fall[1]

1 Virology Department, Institut Pasteur de Dakar, Senegal

2 Kenya Medical Research Institute–Wellcome Trust Research Programme, Kilifi, Kenya

3 Epicentre, Paris, France

*Address all correspondence to: moussa.dia@pasteur.sn

## References


[1] Monath TP, Vasconcelos PFC. Yellow fever. Journal of Clinical Virology. 2015;**64**:160-173

[2] Jonker EFF, Visser LG, Roukens AH. Advances and controversies in yellow fever vaccination. Therapeutic Advances in Vaccines and Immunotherapy. 2013;**1**:144-152

[3] Julander JG, Trent DW, Monath TP. Immune correlates of protection against yellow fever determined by passive immunization and challenge in the hamster model. Vaccine. 2011;**29**:6008-6016

[4] Wieten RW, Jonker EFF, van Leeuwen EMM, et al. A single 17D yellow fever vaccination provides lifelong immunity; characterization of yellow-fever-specific neutralizing antibody and T-cell responses after vaccination. PLoS One. 2016;**11**:e0149871

[5] Horbach IS, de Souza AA, Schwarcz WD, et al. Plaque reduction neutralization test (PRNT) accuracy in evaluating humoral immune response to SARS-CoV-2. Diseases. 2024;**12**:29

[6] Lee WT, Girardin RC, Dupuis AP, et al. Neutralizing antibody responses in COVID-19 convalescent sera. The Journal of Infectious Diseases. 2021;**223**:47-55

[7] Simões M, Camacho LAB, Yamamura AMY, et al. Evaluation of accuracy and reliability of the plaque reduction neutralization test (micro-PRNT) in detection of yellow fever virus antibodies. Biologicals. 2012;**40**:399-404

[8] Roukens AHE, Visser LG. Fractional-dose yellow fever vaccination: An expert review. Journal of Travel Medicine. 2019;**26**:taz024

[9] Casey RM, Harris JB, Ahuka-Mundeke S, et al. Immunogenicity of fractional-dose vaccine during a yellow fever outbreak—Final report. The New England Journal of Medicine. 2019;**381**:444-454

[10] Juan-Giner A, Kimathi D, Grantz KH, et al. Immunogenicity and safety of fractional doses of yellow fever vaccines: A randomised, double-blind, non-inferiority trial. Lancet. 2021;**397**:119-127

[11] Kimathi D, Juan-Giner A, Bob NS, et al. Low-dose yellow fever vaccine in adults in Africa. The New England Journal of Medicine. 2025;**392**:788-797

[12] World Health Organization. WHO position paper on yellow fever vaccine: June 2017. Weekly Epidemiological Record. 2017;**92**(25):269-283

[13] Choi MJ, Heo JY, Seo YB, et al. Six-month longitudinal immune kinetics after mRNA-1273 vaccination: Correlation of peak antibody response with long-term, cross-reactive immunity. Frontiers in Immunology. 2022;**13**:1035441

[14] Dia M, Bob NS, Talla C, et al. Performance assessment and validation of a plaque reduction neutralization test (PRNT) in support to yellow fever diagnostic and vaccine clinical trials. Journal of Medical Virology. 2023;**95**:e28700

[15] Juan-Giner A, Namulwana ML, Kimathi D, et al. Immunogenicity and safety of fractional doses of 17D-213 yellow fever vaccine in children (YEFE): A randomised, double-blind, non-inferiority substudy of a phase 4 trial. The Lancet Infectious Diseases. 2023;**23**:965-973

[16] Kimathi D, Juan-Giner A, Orindi B, et al. Immunogenicity and safety of fractional doses of 17D-213 yellow fever vaccine in HIV-infected people in Kenya (YEFE): A randomised, double-blind, non-inferiority substudy of a phase 4 trial. The Lancet Infectious Diseases. 2023;**23**:974-982

[17] Chong HY, Leow CY, Abdul Majeed AB, et al. Flavivirus infection-a review of immunopathogenesis, immunological response, and immunodiagnosis. Virus Research. 2019;**274**:197770

[18] Dengue—Global situation. Available from: https://www.who.int/emergencies/disease-outbreak-news/item/2024-DON518 [Accessed: 9 April 2025]

[19] Rabe IB, Staples JE, Villanueva J, et al. Interim guidance for interpretation of Zika virus antibody test results. MMWR. Morbidity and Mortality Weekly Report. 2016;**65**:543-546

[20] Doshi RH, Mukadi PK, Casey RM, et al. Immunological response to fractional-dose yellow fever vaccine administered during an outbreak in Kinshasa, Democratic Republic of the Congo: Results 5 years after vaccination from a prospective cohort study. The Lancet Infectious Diseases. 2024;**24**:611-618

[21] Postnikova EN, Pettitt J, Van Ryn CJ, et al. Scalable, semi-automated fluorescence reduction neutralization assay for qualitative assessment of Ebola virus-neutralizing antibodies in human clinical samples. PLoS One. 2019;**14**:e0221407

[22] Rasulova M, Vercruysse T, Paulissen J, et al. A high-throughput yellow fever neutralization assay. Microbiology Spectrum. 2022;**10**:e0254821

[23] Zhang S, Ma P, Orzechowski M, et al. High-throughput neutralization and serology assays reveal correlated but highly variable humoral immune responses in a large population of individuals infected with SARS-CoV-2 in the US between march and august 2020. MBio. 2023;**14**:e0352322

Chapter 6

# Impact of the Obesity in the Clinical Severity Caused by Infection with Orthoflavivirus, Dengue as the Primary Case of Study

*Octavio Rodríguez-Cortés, Ángel Adrián Jiménez-Vega, Mauricio Vázquez-Pichardo, Daniel Duran-Ayala and Raúl Flores-Mejía*

**Abstract**

Orthoflavivirus causes infections that present a significant public health concern in many countries. In the Americas, diseases such as dengue have recently gained attention, with epidemiological alerts issued by the WHO. Many countries in the region have seen a rising trend in overweight and obesity, which increased from 44.4% of adults in the 1990s to 67.5% in 2022. The prevalence of overweight and obesity is expected to reach 73.2% by 2030. Studies have shown that obese individuals have a higher risk (OR 1.37) of developing complications from dengue virus (DENV) infection. Furthermore, obesity and its related complications have been linked to worse outcomes in other infectious diseases, leading to more severe disease courses and higher mortality rates. This chapter examines the main complications associated with obesity-related comorbidities in DENV infection and explores the comorbidities and immune system alterations in individuals with obesity and its comorbidities.

**Keywords:** Orthoflavivirus, dengue, obesity, diabetes, immune response

## 1. Introduction

Obesity is a public health challenge in several countries. In 2022, the WHO estimated that at least one in eight people worldwide were living with obesity, a condition that predisposes to the development of other diseases such as hypertension and diabetes. The prevalence of the latter in 2024 was estimated to be 589 million people between the ages of 20 and 79 years. Over 90% of these cases corresponded to type 2 diabetes (T2D). The major factors contributing to the development of T2D are aging, urbanization, physical inactivity, and the increase in overweight and obesity.

Obesity and T2D have been recognized as independent factors that increase the risk of death and susceptibility to infections. There is solid clinical evidence indicating that the presence of T2D and/or obesity in infections by the influenza virus or SARS-CoV-2 negatively modifies the progression and clinical outcome of the infection. However, the relationship between obesity and other non-respiratory viral infections, as produced by Orthoflavivirus, has been less explored.

Dengue is an epidemiologically relevant disease in tropical and subtropical countries due to the number of cases. Although it commonly presents as asymptomatic, a number of severe cases can lead to death. Obesity and its associated comorbidities (T2D) are factors that affect the severity of the patient's clinical picture.

Therefore, this chapter aims to focus on reviewing the clinical evidence pointing to the consequences derived from obesity and comorbidities in the progression and outcome of infections by Orthoflavivirus, particularly dengue virus (DENV) infection. This chapter also seeks to summarize the information available on the possible immune dysfunction in the antiviral response induced by obesity and T2D and how the presence of these comorbidities likely increases the inflammatory response, promoting the severe stage of dengue. Because these viruses pose a significant threat to public health and frequently impact regions where metabolic diseases are increasingly prevalent, it is essential to understand how these factors interact to improve the clinical management and prevention of complications.

## 2. Clinical course of infections by Orthoflavivirus in people with obesity and associated comorbidities

The major human infections by Orthoflavivirus are those caused by dengue virus (DENV), Zika virus (ZIKV), West Nile virus (WNV), and yellow fever virus (YFV). It is estimated that at least 400 million people present infections related to Orthoflavivirus, mostly caused by DENV. This constitutes an important health risk, especially when considering other factors, such as the number of people living in tropical and subtropical regions, the rise in cases reported in urban areas, the lack of vaccines [1], and the fact that a large number of persons exposed to the infection suffer obesity and comorbidities as T2D, known to modify the clinical progression and outcome of diseases [2].

### 2.1 Dengue virus

A cohort study on infection by DENV and obesity, carried out from 2016 to 2017 in Asian population, reported that in a sample of 173 adults, the risk of presenting severe dengue (SDen) rose by 17% per kilogram of weight that increased the body mass index (BMI), starting at 25 kg/m$^2$. According to its definition, SDen involves the presence of at least one of the following conditions: severe edema leading to shock or liquid accumulation and shortness of breath, alanine transaminase or aspartate transaminase ≥1000 U/L, altered consciousness, and heart failure. Additionally, the risk increases to 37% in the first days of manifestation of the disease. Furthermore, the authors found a significant drop in platelet count among obese patients compared to patients with normal weight and overweight. This is a relevant finding since the mean was well below 150,000 platelets/μL in patients with SDen and obesity, leading to the common handling complications corresponding to this thrombocytopenia [3].

| Year | Author(s) | Type of study | Patients/studies | Country/region | Comorbidity | OR | Main complications/clinical findings |
|---|---|---|---|---|---|---|---|
| 2015 | [6] | Systematic review and meta-analysis | 5 | Asia | T2D | 1.75 (CI 95%: 1.08–2.84) | Increased risk of dengue hemorrhagic fever and shock |
| 2018 | [8] | Systematic review and meta-analysis | 15 | Asia | Obesity | 1.38 (CI 95%: N.R.) | Increased incidence of severe dengue |
| 2018 | [4] | Observational retrospective | 335 | Malaysia | Overweight/obesity | 1.99 (CI 95%: 1.13–3.49) | Fever and elevated hematocrit, increased creatinine and ALT; hospitalization >3 days |
| 2021 | [3] | Cohort prospective | 173 | Malaysia | Obesity | 1.17 (CI 95%: 1.04–1.34) | Association with severe dengue and longer hospitalization |
| 2022 | [5] | Observational retrospective | 1417 | Taiwán | Overweight/obesity | 1.38 (CI 95%: 1.015–1.876) | Petechiae, hepatitis, elevated hematocrit, and decreased platelets |
| 2023 | [9]. | Systematic review and meta-analysis | 6508 | Asia | Overweight/obesity | 1.50 (CI 95%: 1.15–1.97) | Petechiae and bleeding |

*Comparison between studies presented in this chapter, considering study type, sample, comorbidities, complications, and OR of SDen. OR = odds ratio; CI = confidence interval; OR and CI values correspond to those presented in each study.*

**Table 1.**
*OR, main complications, and clinical findings in patients with infection by DENV and obesity.*

Other clinical findings in studies mostly using the adult population report that persons with obesity show kidney alterations, increased creatinine, and longer hospital stays when compared to those with normal weight. The results underline the need for the careful follow-up of this group of patients [4, 5].

Studies report that obesity comorbidities, such as T2D, hypertension, and dyslipidemia, are risk factors for the development of SDen. A systematic review reported that T2D had an odds ratio (OR) of 1.75. On the other hand, a report on cases and controls found ORs of 4.75, 2.77, and 3.65 for the development of SDen in the presence of T2D, hypertension, and dyslipidemia, respectively. The authors found that the OR corresponding to organ failure in SDen was 1.42 and 3.92 in the presence of one or more comorbidities, respectively [6, 7].

A systematic review and meta-analysis of 15 reviews on underage patients published between 2000 and 2016 revealed that obesity has an OR of 1.38 for developing a clinical variation of SDen [8]. Another systematic review and meta-analysis of 6508 pediatric patients, covering articles published between 2000 and 2022, found that children with overweight and obesity had an OR of 1.5 to develop SDen [9]. More attention must be paid to this condition, where complications are developed during infection with SDen among younger patients [10].

### 2.2 Other Orthoflavivirus

Systematic reviews and meta-analyses of studies on WNV have observed that common complications are meningitis, meningoencephalitis, acute encephalitis, respiratory failure, paralysis, and admission into the intensive care unit. These are related to comorbidities, such as diabetes, hypertension, and heart diseases, among which obesity is a predisposing factor. It has been reported that the ORs corresponding to T2D, hypertension, and heart conditions to develop severe disease are 4.21, 2.72, and 6.67, respectively [11].

Studies on diseases caused by the Japanese encephalitis virus (JEV) and ZIKV are not conclusive in terms of complication development due to obesity, T2D, hypertension, or dyslipidemia.

In general, all the studies reported were carried out in the adult Asian population. Then, there is a need for studying other geographic regions and different populations where dengue and other infections associated with Orthoflavivirus are considered a health emergency, mainly in tropical and subtropical regions of Africa and the Americas [12]. On the other hand, evidence from different studies shows that obesity and its major comorbidities must be taken into account by healthcare professionals to improve the care of those infected by Orthoflavivirus, especially DENV (**Table 1**).

## 3. Pathogenesis and immune response against dengue

During cellular immune response to infections by DENV, $CD8^{+}$ T lymphocytes are responsible for the apoptosis of infected cells through a perforin- and granzyme-mediated mechanism. These effector lymphocytes are responsible for recognizing infected cells on the surface of target cells. Additionally, they produce other pro-inflammatory cytokines as IFN-$\gamma$ and TNF-$\alpha$, promoting an increase in local inflammation [13].

The presence of controlled levels of TNF-$\alpha$ can induce the death of infected endothelial cells, supporting the mitigation of the infection. Furthermore, IFN-$\gamma$ also induces HLA class I in infected cells, which likely promotes the presence of effector $CD8^+$ T cells and viral antigen. The amount of infected cells can positively affect inflammatory response regulation. Once the stimulus is finished by the large population of specific clones, the epitopes produced by DENV infection are not required. The action by TNF-α and IFN-γ can lead to increased vascular permeability, plasma extravasation, and hemoconcentration, common complications of SDen [14].

The population of effector $CD8^+$ T cells starts to induce apoptosis, promoted by the end of the stimulus, except memory cells [15] that are kept inactive until a new DENV infection arises.

It is known that IgM antibodies are involved in the early identification of external invasion and the elimination of pathogens as bacteria and viruses. In SDen, the IgM complex has been consistently found in the walls of skin, taste buds, and skin rash blood vessels. Additionally, there is an increase in platelet-associated IgM. The role of the IgM complex circulating in SDen remains partially unknown. The IgM complex adhered to the platelet surface can probably increase platelet destruction by the reticuloendothelial system in the liver and spleen, promoting thrombocytopenia in SDen. Therefore, IgM levels specific for DENV may affect the outcome of infection by DENV. These effects can also be induced by IgG in a secondary infection by DENV.

There are sufficient high-affinity antibodies aimed at epitopes in virus-cell fusion peptides, located at domain II of the envelope protein (E) during infection by DENV, which can promote the neutralization of viral particles [16]. However, dengue antibodies with low avidity or at suboptimal concentrations can bind to the viral particle under subneutralizing conditions [17]. In both scenarios, these sub-neutralizing or non-neutralizing antibodies can lead to antibody-dependent potentiation (antibody-dependent enhancement, ADE), where antibodies bound on the viral surface are recognized by Fc receptors in target cells. Once they are recognized, especially by APC, phagocytosis of the viral particle is induced, leading to an increase in the number of infected cells and, in consequence, greater viremia. This phenomenon might contribute to the development of clinical symptoms typical of severe cases and linked to increased viremia [18].

Although ADE and cytokine storm are well-accepted phenomena that have been studied for decades, it must be highlighted that most of the cases are solved without any signs of disease. There are more pieces to the DENV vs. immune system puzzle, and it seems that cases in which the immune response leads to a severe disease are also related to factors such as preexisting immunity; interval between infections; the host's genetics; viral serotype; age; and comorbidities as obesity, hypertension, and diabetes. These must be considered among the factors that affect the severity of infection by DENV [19].

The pathogenesis of dengue is characterized by a hyperactivation of immune cells that ends in cytokine production. This, along with other factors, can be the key to defining the course of the disease.

Differences in cytokine levels in SDen patients have been analyzed, patients with SDen have reported higher levels of IL-13 and IL-14. Increased IL-10 has already been proposed as an indicator of dengue disease severity [20]. Among secondary infections, cases that did not progress to SDen showed higher levels of IL-10 than those in non-severe cases of dengue. It is known that IL-17 is associated with pleural effusion and respiratory issues both in children and adults. In cases of SDen, patients had a high number of TH17 cells in kidney tissue, contributing to intense, acute inflammation and leading to tissue injury, promoting increased vascular permeability [21]. Most of the studies demonstrated higher IL-18 in SDen patients vs. those with non-severe dengue [22–24].

High levels of IFN-γ have been found in patients infected with different DENV serotypes [25–27], and they were even higher in people who presented the most severe manifestations at an early stage [28–30]. Still, Chen et al. (2007) and Priyadarshini et al. (2010) found higher IFN-γ in patients with non-severe dengue vs. patients with SDen [31, 32].

Patients with SDen presented higher TNF-α as compared to those who suffered non-severe dengue [33, 34].

Given the multifactorial infection by DEV and the immunopathogenesis of the disease, the course should not only be considered as a response to one or a few aspects of the host's immune response during infection by DENV. Only a comprehensive view of topics related to infection, local and systemic response, viral factors, comorbidities, host's immunogenetic background, and epidemiological panorama of cocirculation of more than one Orthoflavivirus may shed light on dengue as a key disease for health systems.

Then, generating constant scientific evidence that provides information to elucidate the processes involving the triggering of severe disease constitutes a challenge.

## 4. Alterations in the immune response among patients with obesity and diabetes and repercussions in infection by DENV

The epidemiological evidence has shown that obesity and diabetes are independent risk factors that increase mortality and susceptibility to infections by different pathogens. Several systematic reviews with meta-analyses have demonstrated that the relative risk (RR) of mortality by infections (excluding pneumonia) in people with diabetes is 2.39 (CI 1.95–3.82) while that of pneumonia is 1.67 (CI 1.45–1.92). In addition, a BMI above 35 increases mortality (RR > 1.5) as compared to a BMI of 22.5 [35, 36]. It has been observed that the presence of T2D increases the RR of developing active tuberculosis to 3.11 (CI 2.27–7.83) [2, 37]. This is fundamental to consider T2D as a state of secondary immunodeficiency.

Solid clinical evidence indicates that the presence of obesity and T2D negatively modifies the course and clinical outcome of influenza and COVID-19, mostly related to inflammatory deregulation. An observational study including almost 200,000 patients confirmed with COVID-19 before vaccination reported through a multivariate analysis that the presence of T2D and obesity increases the probability of death with ORs of 2.09 (CI 2.30–2.16) and 1.43 (CI 1.38–1.48), respectively.

This epidemiological evidence associating negative prognosis and clinical outcomes of respiratory viral infections in people with obesity and T2D is explained by many situations. One of the most important involves the alteration in the immune and inflammatory states related to obesity and T2D. These comorbidities can result in antiviral response dysfunction and an exacerbation of the pro-inflammatory response, leading to organ failure.

The state of the immune system is characterized by presenting chronic low-grade inflammation (CLGI) or meta-inflammation in patients with obesity and diabetes. This state is caused by interactions between the immune system and products of metabolic alterations associated with obesity and diabetes.

As a result of obesity in visceral adipose tissue (VAT), the excess of soluble mediators, such as hormones, cytokines, and adipokines, is released. During obesity, VAT is commonly scarcely vascularized, producing hypoxia and adipose hypertrophy. This leads to necrosis and the release of triglycerides (TGs), free fatty acids (FFAs), and danger-associated molecular patterns (DAMPs). Together, these soluble stimuli act upon macrophages and tissue T lymphocytes, which, in response, induce the release of inflammatory mediators, such as TNF-α, IL-1β, IL-6, IL-8, and IL-12, among others, that exacerbate and alter CLGI. When the individual shows more comorbidities, such as hypertension, dyslipidemia, or T2D, LDL and AGEs are also added to the list of CLGI-inducing stimuli. This state can potentially lead to both an exacerbated inflammatory response and, paradoxically, a suboptimal immune system response (**Figure 1**).

In the particular case of dengue infection, the main clinical characteristic of SDen is severe plasma leakage, along with organ failure and thrombocytopenia. As previously described, the presence of obesity increases the likelihood of developing SDen both in children and adults. Next, we will discuss how CLGI linked to obesity and T2D might promote this clinical variant of dengue with a negative prognosis.

Although an integral immune response is necessary for the protection against Orthoflaviviruses, the production of type I IFNs, the activation of natural killer (NK) cells, the production of neutralizing antibodies, and the profiles of Th1 and Th17 lymphocytes are key to protection against dengue, as in all antiviral responses.

Studies in persons with obesity and animal models *in vitro* indicate that several soluble signals related to obesity, such as palmitic acid, AGL, or TNF-α (through receptors), promote the production of IFN-β and pro-inflammatory cytokines from the NF-kB pathway. Furthermore, during obesity, there is mitochondrial dysfunction, which induces mitochondrial DNA release, activating the STING (Stimulator of Interferon Genes) pathway for IFN-β production. So, the production of this cytokine

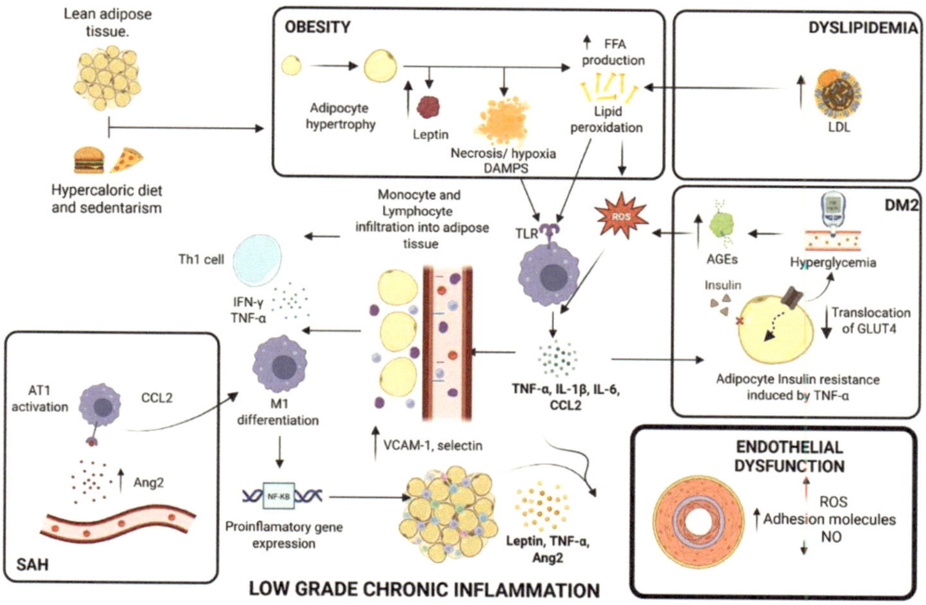


**Figure 1.**
*Pathophysiology of chronic low-grade inflammation. Induction of a chronic inflammatory state, resulting from metabolic and cellular alterations caused by obesity and other metabolic disorders, in addition to alterations in the vascular endothelium, secondary to this state. FFA: free fatty acids; LDL: low-density lipoprotein; DAMPS: damage-associated molecular patterns; AGEs: advanced glycation end-products; ROS: reactive oxygen species; TLR: toll-like receptor; Glut4: glucose transporter type 4; TNF-α: tumor necrosis factor-alpha; IL: interleukin; CCL2: monocyte chemoattractant Protein-1; IFN-γ: interferon gamma; Ang2: angiotensin 2; NO: nitric oxide. Created in https://BioRender.com*

with antiviral activity is increased under obesity conditions [38], which would help in the induction of the anti-replicative state for better viral control. Still, the increase in TNF-α, IL-β, and IL-6 of CLGI associated with obesity and T2D increases the physio-pathological processes that characterize SDen, as cytokine storm, and severe plasma leakage that leads to multiple organ failure.

In obesity, CLGI disturbs vascular homeostasis and leads to its dysfunction. Additionally, the inflammation of visceral adipose and perivascular tissues leads to a reduction in nitric oxide (NO) bioavailability, while other components of metabolic syndrome, as insulin resistance (IR) and oxidized low-density lipoproteins (oxLDL), also promote endothelial dysfunction. When infected by DENV, dysfunctional endothelium is more prone to disruption and plasma leakage, a characteristic of SDen.

Human NK cells express receptors for adipokines, IL-6, leptin, and adiponectin, so they can respond to alterations in levels of these soluble signals, as those observed in obesity. Studies on the effect of obesity on the function of human peripheral blood NK cells focus on the impact on cytotoxic function and IFN-γ production. In that regard, most of the studies report nonsignificant alterations in the production of IFN-γ and granzymes and their decreased cytotoxicity. Particularly, adipokines, IL-6, and leptin are increased in obesity, and although no association between plasma concentration of IL-6 and leptin and NK cells in blood has been found, there is a relationship with modifications to their effector functions. A reduced expression of the cytotoxic proteins perforin and granzyme, present in NK granules, is induced by IL-6. However, there are conflicting results on the effect of IL-6 and leptin on

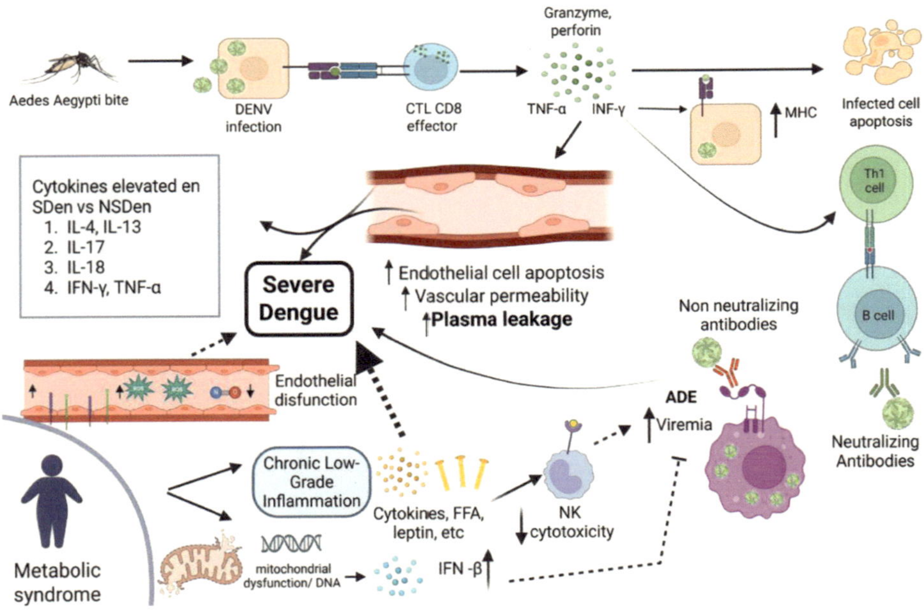


**Figure 2.**
*Influence of chronic low-grade inflammation in the pathophysiology of SDen. Mechanisms of immune response against DENV infection and the relationship of chronic low-grade inflammation in the development of SDen, as well as the main elevated cytokines in obesity and SDen. CTL: Cytolytic T lymphocyte; NSDen: No severe dengue; TNF-α: Tumor necrosis factor-alpha; IFN-γ: Interferon gamma; ROS: Reactive oxygen species; FFA: Free fatty acids; IFN-β: Interferon-beta; IL: Interleukin; ADE: antibody-dependent enhancement; NK: Natural killer; MHC: Major histocompatibility complex. Created in https://BioRender.com*

cytotoxicity and IFN-γ release, as these functions have been reported to increase, decrease, and remain unchanged [39]; that is, a higher concentration of IL-6 linked to CLGI in persons with obesity reduced perforin and granzyme synthesis in human NK cells, which might lead to reduced containment of DENV.

Regarding the adaptive response in infection against DENV, the production of IgG antibodies is key to the neutralization of the virus. Still, in the secondary response, the antibodies can also take part in ADE, which increases infection and pro-inflammatory state by monocyte and macrophage activation (**Figure 2**).

In that sense, people with obesity and T2D show increased populations of Th1 and Th17 lymphocytes and a reduction in regulatory T cells in peripheral blood [40]. Through the production of IFN-γ, Th1 cells can contribute to IgG antibody production and the activation of monocytes and macrophages. Infection with DENV also promotes ADE and cytokine storm, both of which are present in SDen [41].

## 5. Future prospects: The need for specific epidemiological studies

As previously described, there are discrepancies and information gaps regarding complications caused by Orthoflavivirus infections in people with obesity and their comorbidities. Some of these discrepancies may be the result of factors not unified in studies, such as considering the number of patients diagnosed with obesity through at least two criteria and excluding overweight patients, disparate age groups, and the number of components of metabolic syndrome from the analysis. It is also necessary to unify the criteria to determine the severity of clinical manifestations of the infection.

## 6. Conclusions

Infections caused by the dengue virus account for the majority of the clinical information on humans, so there is a need for studying other diseases caused by epidemiologically relevant Orthoflavivirus.

The scientific evidence indicates that obesity, type 2 diabetes, and hypertension are risk factors for developing complications during the course of the diseases, and they must be considered by healthcare professionals to improve attention to those infected by Orthoflavivirus, mostly the dengue virus.

## Acknowledgements

The authors acknowledge the financial support received from the Secretary of Research and Postgraduate Studies (SIP 20250195, 20250302, and 20253561) of the National Polytechnic Institute (IPN), which facilitated the development of research projects and the publication of scientific texts.

## Conflict of interest

The authors declare no conflict of interest.

## Author details

Octavio Rodríguez-Cortés[1], Ángel Adrián Jiménez-Vega[1], Mauricio Vázquez-Pichardo[2], Daniel Duran-Ayala[2] and Raúl Flores-Mejía[1*]

1 Laboratorio de Inflamación y Obesidad, Escuela Superior de Medicina, Instituto Politécnico Nacional, Ciudad de México, México

2 Laboratorio de Arbovirus, Instituto de Diagnóstico y Referencia Epidemiológicos (InDRE) "Dr. Manuel Martínez Báez", Secretaría de Salud, Ciudad de México, México

*Address all correspondence to: raflores@ipn.mx

## References

[1] Pierson TC, Diamond MS. The continued threat of emerging flaviviruses. Nature Microbiology. 2020;**5**(6):796-812. DOI: 10.1038/s41564-020-0714-0. Epub 2020 May 4

[2] León J, Rodríguez O, Flores R, Gaona C, Gonzalez A. Impact of metabolic syndrome in the clinical outcome of disease by SARS-COV-2. Archives of Medical Research. 2021;**52**(7):738-745. DOI: 10.1016/j.arcmed.2021.04.001

[3] Zulkipli MS, Rampal S, Bulgiba A, Peramalah D, Jamil N, See LLC, et al. Is there any association between body mass index and severity of dengue infection? Transactions of the Royal Society of Tropical Medicine and Hygiene. 2021;**115**(7):764-771. DOI: 10.1093/trstmh/trab021

[4] Tan VPK, Ngim CF, Lee EZ, Ramadas A, Pong LY, Ng JI, et al. The association between obesity and dengue virus (DENV) infection in hospitalised patients. PLoS One. 2018;**13**(7):e0200698. DOI: 10.1371/journal.pone.0200698

[5] Chiu YY, Lin CY, Yu LS, Wang WH, Huang CH, Chen YH. The association of obesity and dengue severity in hospitalized adult patients. Journal of Microbiology, Immunology, and Infection. 2023;**56**(2):267-273. DOI: 10.1016/j.jmii.2022.08.008. Epub 2022 Aug 18

[6] Htun NSN, Odermatt P, Eze IC, Boillat-Blanco N, D'Acremont V, Probst-Hensch N. Is diabetes a risk factor for a severe clinical presentation of dengue? - review and meta-analysis. PLoS Neglected Tropical Diseases. 2015;**9**(4):e0003741. DOI: 10.1371/journal.pntd.0003741

[7] Pang J, Hsu JP, Yeo TW, Leo YS, Lye DC. Diabetes, cardiac disorders and asthma as risk factors for severe organ involvement among adult dengue patients: A matched case-control study. Scientific Reports. 2017;7:39872. DOI: 10.1038/srep39872

[8] Zulkipli MS, Dahlui M, Jamil N, Peramalah D, Wai HVC, Bulgiba A, et al. The association between obesity and dengue severity among pediatric patients: A systematic review and meta-analysis. PLoS Neglected Tropical Diseases. 2018;**12**(2):e0006263. DOI: 10.1371/journal. Pntd.0006263

[9] Chen CY, Chiu YY, Chen YC, Huang CH, Wang WH, Chen YH, et al. Obesity as a clinical predictor for severe manifestation of dengue: A systematic review and meta-analysis. BMC Infectious Diseases. 2023;**23**(1):502. DOI: 10.1186/s12879-023-08481-9

[10] Verhagen LM, de Groot R. Dengue in children. The Journal of Infection. 2014;**69**(Suppl 1):S77-S86. DOI: 10.1016/j.jinf.2014.07.020. Epub 2014 Sep 13

[11] Badawi A, Velummailum R, Ryoo SG, Senthinathan A, Yaghoubi S, Vasileva D, et al. Prevalence of chronic comorbidities in dengue fever and West Nile virus: A systematic review and meta-analysis. PLoS One. 2018;**13**(7):e0200200. DOI: 10.1371/journal. Pone.0200200

[12] Hou B, Chen H, Gao N, An J. Cross-reactive immunity among five medically important mosquito-borne flaviviruses related to human diseases. Viruses. 2022;**14**:1213. DOI: 10.3390/v14061213

[13] Mori M, Kurane I, Janus J, Ennis FA. Cytokine production by dengue virus

antigen-responsive human T lymphocytes in vitro examined using a double immunocytochemical technique. Journal of Leukocyte Biology. 1997;**61**(3):338-345. DOI: 10.1002/jlb.61.3.338

[14] Leong AS, Wong KT, Leong TY, Tan PH, Wannakrairot P. The pathology of dengue hemorrhagic fever. Seminars in Diagnostic Pathology. 2007;**24**(4):227-236. DOI: 10.1053/j.semdp.2007.07.002

[15] Sun P, Fernandez S, Marovich MA, et al. Functional characterization of ex vivo blood myeloid and plasmacytoid dendritic cells after infection with dengue virus. Virology. 2009;**383**(2):207-215. DOI: 10.1016/j.virol.2008.10.022

[16] Seligman SJ, Gould EA. Safety concerns with regard to live attenuated flavivirus vaccines. The Journal of Infectious Diseases. 2008;**198**(5):794-795. DOI: 10.1086/590915

[17] Halstead S. Antibodies determine virulence in dengue. Annals of the New York Academy of Sciences. 2009;**1171**(1):E48-E56. DOI: 10.1111/j.1749-6632.2009.05052.x

[18] Vaughn DW, Green S, Kalayanarooj S, Innis BL, Nimmannitya S, Suntayakorn S, et al. Dengue viremia titer, antibody response pattern, and virus serotype correlate with disease severity. The Journal of Infectious Diseases. 2000;**181**(1):2-9. DOI: 10.1086/315215

[19] Guzman MG, Gubler DJ, Izquierdo A, Martinez E, Halstead SB. Dengue infection. Nature Reviews. Disease Primers. 2016;**2**:16055. DOI: 10.1038/nrdp.2016.55

[20] Tauseef A, Umar N, Sabir S, Akmal A, Sajjad S, Zulfiqar S. Interleukin-10 as a marker of disease progression in dengue hemorrhagic fever. Journal of the College of Physicians and Surgeons–Pakistan. 2016;**26**(3):187-190

[21] Pagliari C, Quaresma J, Kanashiro-Galo L, Carvalho L, Vitoria W, da Silva W, et al. Human kidney damage in fatal dengue hemorrhagic fever results of glomeruli injury mainly induced by IL17. Journal of Clinical Virology. 2016;**75**:16-20V. ISSN 1386-6532. DOI: 10.1016/j.jcv.2015.12.005

[22] Diamond M, Edgil D, Roberts T, Lu B, Harris E. Infection of human cells by dengue virus is modulated by different cell types and viral strains. Journal of Virology. 2000;**74**(17):7814-7823. DOI: 10.1128/JVI.74.17.7814-7823.2000

[23] Azeredo E, Zagne S, Alvarenga A, Nogueira R, Kubelka C, de Oliveira-Pinto L. Activated peripheral lymphocytes with increased expression of cell adhesion molecules and cytotoxic markers are associated with dengue fever disease. Memórias do Instituto Oswaldo Cruz. 2006;**101**(4):437-449. DOI: 10.1590/S0074-02762006000400016

[24] Van Wilgenburg B, Scherwitzl I, Hutchinson E, Leng T, Kurioka A, Kulicke C, et al. MAIT cells are activated during human viral infections. Nature Communications. 2016;**7**:11653. DOI: 10.1038/ncomms11653

[25] Soundravally R, Hoti S, Patil S, Cleetus C, Zachariah B, Kadhiravan T, et al. Association between proinflammatory cytokines and lipid peroxidation in patients with severe dengue disease around defervescence. International Journal of Infectious Diseases. 2014;**18**:68-73. DOI: 10.1016/j.ijid.2013.09.022

[26] Cruz S, Puerta H, Flores H, González S, López I, Ortiz V, et al.

Primary dengue virus infections induce differential cytokine production in Mexican patients. Memórias do Instituto Oswaldo Cruz. 2016;**111**(3):161-170. DOI: 10.1590/0074-02760150359

[27] Feitosa R, Vallinoto A, Vasconcelos P, Azevedo R, Azevedo V, Machado L, et al. Gene polymorphisms and serum levels of pro- and anti-inflammatory markers in dengue viral infections. Viral Immunology. 2016;**29**(7):379-388. DOI: 10.1089/vim.2016.0026

[28] Liao B, Tang Y, Hu F, Zhou W, Yao X, Hong W, et al. Serum levels of soluble vascular cell adhesion molecules may correlate with the severity of dengue virus-1 infection in adults. Emerging Microbes & Infections. 2015;**4**(4):e24. DOI: 10.1038/emi.2015

[29] Cui L, Lee Y, Thein T, Fang J, Pang J, Ooi E, et al. Serum metabolomics reveals serotonin as a predictor of severe dengue in the early phase of dengue fever. PLoS Neglected Tropical Diseases. 2016;**10**(4):e0004607. DOI: 10.1371/journal.pntd.0004607

[30] Zhao H, Zhang F, Zhu Q, Wang J, Hong W, Zhao L, et al. Epidemiological and virological characterizations of the 2014 dengue outbreak in Guangzhou, China. PLoS One. 2016;**11**(6):e0156548. DOI: 10.1371/journal.pone.0156548

[31] Priyadarshini D, Gadia R, Tripathy A, Gurukumar R, Bhagat A, Patwardhan S, et al. Clinical findings and pro-inflammatory cytokines in dengue patients in Western India: A facility-based study. PLoS One. 2010;**5**(1):e8709. DOI: 10.1371/journal.pone.0008709

[32] Chen R, Yang K, Wang L, Liu J, Chiu C, Cheng J. Different clinical and laboratory manifestations between dengue haemorrhagic fever and dengue fever with bleeding tendency. Transactions of the Royal Society of Tropical Medicine and Hygiene. 2007;**101**(11):1106-1113. DOI: 10.1016/j.trstmh.2007.06.019

[33] Kittigul L, Temprom W, Sujirarat D, Kittigul C. Determination of tumor necrosis factor-alpha levels in dengue virus infected patients by sensitive biotin-streptavidin enzyme-linked immunosorbent assay. Journal of Virological Methods. 2000;**90**(1):51-57. DOI: 10.1016/S0166-0934(00)00215-9

[34] Villar L, Gélvez R, Rodríguez J, Salgado D, Parra B, Osorio L, et al. Biomarcadores pronósticos de gravedad del dengue. Biomédica. 2013;**33**(suppl.1):108-116. DOI: 10.7705/biomedica.v33i0.733

[35] Seshasai R, Kaptoge S, Thompson A, Di Angelantonio E, Gao P, Sarwar N, et al. Emerging risk factors collaboration diabetes mellitus, fasting glucose, and risk of cause-specific death. New England Journal of Medicine. 2011;**369**(9):829-841. DOI: 10.1056/NEJMoa1008862

[36] Lennon H, Sperrin M, Badrick E, Renehan A. The obesity paradox in cancer: A review. Current Oncology Reports. 2016;**18**(9):56. DOI: 10.1007/s11912-016-0539-4

[37] Jeon C, Murray M. Diabetes mellitus increases the risk of active tuberculosis: A systematic review of 13 observational studies. PLoS Medicine. 2008;**5**(7):e152. DOI: 10.1371/journal.pmed.0050152

[38] Gong J, Gao X, Ge S, Li H, Wang R, Zhao L. The role of cGAS-STING signalling in metabolic diseases: From signalling networks to targeted intervention. International Journal of Biological Sciences. 2024;**20**(1):152-174. DOI: 10.7150/ijbs.84890

[39] Bähr I, Spielmann J, Quandt D, Kielstein H. Obesity-associated alterations of natural killer cells and immunosurveillance of cancer. Frontiers in Immunology. 2020;**11**:245. DOI: 10.3389/fimmu.2020.00245

[40] Engin A. Endothelial dysfunction in obesity and therapeutic targets. Advances in Experimental Medicine and Biology. 2024;**1460**:489-538. DOI: 10.1007/978-3-031-63657-8_17

[41] Touch S, Clément K, André S. T cell populations and functions are altered in human obesity and type 2 diabetes. Current Diabetes Reports. 2017;**17**(9):81. DOI: 10.1007/s11892-017-0900-5

Chapter 7

# Liver Damage Caused by Dengue

*Jazmín García-Machorro, Gabriela Mellado-Sánchez and José Antonio Morales-González*

## Abstract

Dengue is an acute viral infection caused by a member of the *Flaviviridae* family, genus *Orthoflaviviridae*, and is usually transmitted by mosquitoes of the genus *Aedes*. Symptoms may include fever, headache, retroocular pain, myalgia, arthralgia, generalized pain, and rash. In some cases, liver damage has been found, ranging from elevated transaminases (ALT, alanine aminotransferase, and AST, aspartate aminotransferase), total bilirubin, steatosis, apoptosis, necrosis, hemorrhage, and even liver failure; therefore, proper diagnosis and management are important. Dengue infection triggers overexpression of pro-inflammatory cytokines, causing structural changes in mitochondria that activate cell death cascades. Additionally, it is recommended to evaluate a history of alcohol consumption (acute or chronic) due to the toxicity it induces in hepatocytes, and reduces the capacity to respond to the virus. In cases of excessive alcohol consumption, it triggers the activation of Kupffer cells and further liver damage caused by the virus.

**Keywords:** dengue virus, DENV infection, hepatic damage, inflammation, apoptosis, death

## 1. Introduction

The *Flaviviridae* family is highly diverse and is characterized by the inclusion of enveloped, positive-sense, single-stranded RNA viruses. It comprises pathogens of medical and veterinary importance, as well as emerging viruses that pose potential threats to public health. The *Flaviviridae* family is divided into four genera: *Orthoflavivirus*, *Pestivirus*, *Pegivirus*, and *Hepacivirus* [1, 2]. The *Orthoflavivirus* genus includes approximately 50 viral species, characterized by transmission cycles involving both vertebrates and invertebrates. These viruses are capable of causing neurotropic or systemic diseases in vertebrate hosts. Several *Orthoflaviviruses* that infect humans originated in sylvatic environments, where they circulate between nonhuman primates and arthropods. Some species evolved to maintain endemic cycles within human communities *via* vectors such as mosquitoes and later adapted to urban environments, as is the case with the dengue virus (DENV) in tropical regions. Others, like the Zika virus (ZIKV) in the Americas, have caused pandemic outbreaks with significant health consequences. Likewise, yellow fever virus (YFV) continues to cause morbidity and mortality in areas where sylvatic cycles persist or where vaccination access is limited [3, 4]. *Orthoflaviviruses* possess a single-stranded, positive-sense RNA genome of approximately 11 kb, enclosed in an enveloped, icosahedral virion approximately 50 nm in diameter. This genus can be classified according to the hosts involved in transmission: mosquito-borne species (*Culex spp.* or *Aedes spp.*), tick-borne species, insect-specific viruses, or those with unidentified invertebrate vectors. Their phylogeny is closely related to the ecological niches of their vectors. For example, *Orthoflaviviruses* transmitted by *Culex spp.* evolved from those transmitted by *Aedes* spp. and circulate primarily in birds, although some mammals can act as reservoirs. In contrast, *Aedes*-borne viruses maintain their cycle in primates and other mammals, with birds sometimes acting as terminal hosts [5].

In the following sections, we will first describe the functions of the liver under normal conditions. Then, we will discuss *Orthoflaviviruses* associated with liver damage, with a focus on DENV: its general characteristics, affected *in vitro* cell models, potential mechanisms of liver injury, and contributing factors such as prior alcohol-induced liver damage.

## 2. Liver

The liver is the largest gland in the human body, weighing approximately 1500 g. It is located in the abdominal cavity, occupying the upper right quadrant and part of the left. It is considered a mixed gland due to its endocrine and exocrine functions. The liver is surrounded by a thin connective tissue capsule known as Glisson's capsule, which is mostly covered by the peritoneum [6].

The liver is a vital organ that performs essential metabolic functions. It synthesizes and secretes many plasma proteins, plays a crucial role in the uptake, storage, and distribution of numerous nutrients and vitamins, maintains blood glucose levels, regulates plasma concentrations of very-low-density lipoproteins (VLDL), detoxifies or conjugates many toxic substances and drugs, and produces bile. Additionally, it plays important roles in endocrine and immune system regulation.

One key feature that enables the liver to fulfill its many functions is its dual blood supply. About 70–75% of the liver's blood supply comes from the hepatic portal vein, with the remainder supplied by the hepatic artery (a branch of the celiac trunk). The portal vein carries venous blood from the gastrointestinal tract, pancreas, and spleen, containing nutrients (except lipids, which travel *via* the lymphatic system), toxins absorbed in the intestine, blood cells and their degradation products, as well as endocrine secretions from the pancreas and enteroendocrine cells. The arterial blood is more oxygenated but contains fewer nutrients than portal blood and is required for the liver's metabolic activity. Blood from both vessels mixes before perfusing the liver cells, passing through sinusoidal capillaries that bathe the hepatocytes. As a result, hepatocytes are the first to encounter nutrients absorbed from the gut and are also the first exposed to toxins. Importantly, they are never exposed to fully oxygenated blood [6].

### 2.1 Liver functions

#### *2.1.1 Metabolic function*

The liver plays a key role in carbohydrate metabolism. Glucose absorbed in the digestive tract is phosphorylated in the liver to form glucose-6-phosphate. Depending on the body's energy demands, glucose-6-phosphate can enter glycolytic pathways or be stored as hepatic glycogen. During fasting, glycogen undergoes glycogenolysis to release glucose into the bloodstream. Gluconeogenesis also occurs in hepatocytes [7].

Hepatocytes take up plasma fatty acids and use them for energy *via* β-oxidation and also produce ketone bodies, which serve as fuel for other organs. Another important function is cholesterol uptake and synthesis for the production of bile salts, VLDL, and cell organelle biosynthesis. Ammonium ions generated from protein and nucleic acid degradation are converted into urea in the liver, which is the primary site of urea production.

### *2.1.2 Plasma protein synthesis*

The liver synthesizes many plasma proteins, including albumin, which regulates blood volume and tissue fluid formation; lipoproteins, primarily VLDL for triacylglycerol transport, and synthesis of smaller amounts of low-density lipoproteins (LDL) delivering cholesterol to tissues and high-density lipoproteins (HDL) transporting cholesterol from tissues to the liver; and implicated in coagulation as prothrombin and fibrinogen, which are essential for the blood coagulation cascade, whereas alpha and beta globulins, besides coagulation, are related to colloid osmotic pressure and substance transport. Additionally, some specific globulins are involved in immune functions, among them alpha-1-globulin and beta-2-microglobulin, which are part of the complement system and immune response; meanwhile, acute-phase proteins are part of the inflammatory response [8].

### *2.1.3 Storage and processing of vitamins and iron*

The liver captures, stores, and regulates levels of vitamin A (retinol) and synthesizes its transport protein. Vitamin D (cholecalciferol), vital for calcium and phosphorus metabolism, is converted in the liver to 25-hydroxycholecalciferol, which is then transformed in the kidneys into its active form, 1,25-dihydroxycholecalciferol. Vitamin K is transported from the intestine to the liver and is involved in synthesizing prothrombin and other coagulation factors. The liver is also central to iron storage, metabolism, and homeostasis. It produces most proteins involved in iron transport and metabolism (e.g., transferrin, haptoglobin, hemopexin). Iron is stored in hepatocyte cytoplasm as ferritin and hemosiderin granules [6, 9].

### *2.1.4 Detoxification function*

Many drugs and xenobiotics are not water-soluble and cannot be excreted *via* the kidneys. Hepatocytes convert these substances into more soluble forms through two phases: oxidation and conjugation.

Oxidation, via hydroxylation or carboxylation in the smooth endoplasmic reticulum or mitochondria, primarily involving cytochrome P450 enzymes.

Conjugation with glucuronic acid, glycine, taurine, or sulfate radicals [6, 10].

### *2.1.5 Ethanol metabolism*

The liver is the primary site for metabolizing most drugs and xenobiotics, including ethanol [11]. After ingestion, 30% of ethanol is absorbed in the stomach and 70% in the duodenum, then distributed through body water. It is almost entirely oxidized in the liver at a zero-order kinetic rate of 1–20 mg/dL/hour, depending on body weight. About 90% is metabolized in the liver, while 10% is excreted *via* extrahepatic routes such as the stomach, kidneys, lungs, and sweat [6, 12].

Ethanol is oxidized to acetaldehyde by three major enzyme systems:

a. Alcohol dehydrogenase (ADH) in the cytoplasm, accounting for 90% of ethanol metabolism. Women have lower ADH activity, resulting in higher blood ethanol levels.

b. Microsomal ethanol oxidizing system (MEOS), a cytochrome P450 2E1–dependent system, responsible for 10% of ethanol oxidation in moderate drinkers but upregulated in chronic drinkers.

c. Catalase system, dependent on hydrogen peroxide (in peroxisomes), with minimal contribution.

All three systems convert ethanol into acetaldehyde, which is then oxidized to acetate by aldehyde dehydrogenase (ALDH) (**Figure 1**). Acetaldehyde is a highly reactive and potentially toxic compound. Under normal conditions, acetate production is favored. However, reduced ALDH activity leads to acetaldehyde accumulation in blood and tissues, causing flushing, tachycardia, and, in severe cases, shock.

Chronic ethanol consumption increases liver ethanol metabolism by up to 30%, inducing metabolic tolerance. Cellular and behavioral tolerance may follow, marked by structural and biochemical liver changes. Upon alcohol withdrawal, these changes may take weeks to reverse, indicating physical dependence [6, 13].

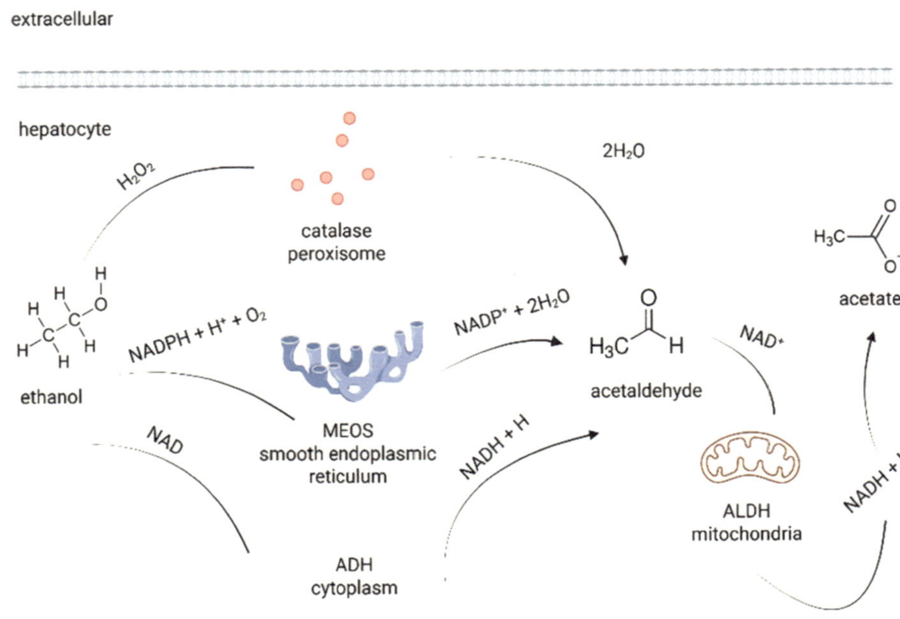


**Figure 1.**
*Ethanol metabolism: Ethanol is oxidized to acetaldehyde and then to acetate. ADH: Alcohol dehydrogenase; MEOS: Microsomal ethanol oxidizing system; ALDH: Aldehyde dehydrogenase.*

### 2.2 Endocrine functions of the liver

The liver can metabolize and modify hormones secreted by other organs. For example, thyroxine (T4), which is secreted by the thyroid gland, undergoes deiodination in the liver to form triiodothyronine (T3), the biologically active form of the hormone.

Additionally, the liver synthesizes growth hormone–releasing factor (GHRF), which modulates the action of growth hormone produced by the pituitary gland [6].

### 2.3 Immune functions of the liver

The liver contains specialized cells with immune functions. Kupffer cells are resident macrophages located within hepatic sinusoids. They are capable of phagocytosing viruses, bacteria, and foreign macromolecules to eliminate pathogens.

Pit cells are liver-specific natural killer (NK) cells with abundant cytotoxic granules that enable them to exert potent cytolytic activity [14].

In addition to immune cells, the liver synthesizes several immune-related proteins, including acute-phase proteins, complement components, and cytokines, which contribute to the immune response and the defense against infections. Thus, hepatocytes are among the first cells to encounter antigens and microorganisms that enter the bloodstream, especially those derived from the gastrointestinal tract [15].

## 3. *Orthoflaviviruses* that affect the liver

Liver functions can be disrupted by multiple factors such as the use of certain medications, exposure to toxins, metabolic and non-metabolic diseases (e.g., congestive heart failure, Reye's syndrome, some autoimmune diseases), excessive alcohol consumption, and viral infections. In this section, we describe viruses that affect the liver, focusing on *Orthoflaviviruses*.

The most well-known hepatotropic viruses are hepatitis viruses A, B, C, D, and E, which belong to different families and genera:

Hepatitis A: *Picornaviridae, Hepatovirus*;
Hepatitis B: *Hepadnaviridae, Orthohepadnavirus*;
Hepatitis C: *Flaviviridae, Hepacivirus*;
Hepatitis D: *Kolmioviridae, Deltavirus*;
Hepatitis E: *Hepeviridae, Orthohepevirus*.

These viruses can cause liver inflammation and severe disease, including hepatocellular carcinoma or death [16–20].

The *Flaviviridae* family includes the *Orthoflavivirus* genus, which comprises DENV, YFV, and ZIKV. These can cause a range of symptoms, from mild (e.g., nausea, vomiting, headache) to severe complications such as hemorrhage, meningitis, microcephaly, and hepatic or renal failure. Hepatic alterations from these viruses range from tissue injury to biochemical abnormalities.

Many Orthoflaviviruses are also considered arboviruses, a term used for viruses transmitted to humans and other animals by arthropods, primarily blood-feeding insects such as ticks and mosquitoes. Arboviruses can infect both invertebrate and vertebrate hosts and can be classified into three categories: tick-borne (e.g., tick-borne encephalitis virus [TBEV], Powassan virus [POWV], and deer tick virus

[DTV]), mosquito-borne (e.g., DENV, ZIKV, YFV, West Nile virus [WNV], and Japanese encephalitis virus [JEV]), and those with unknown vectors [21, 22].

Despite the public health relevance of research on Orthoflaviviruses, understanding their pathophysiology remains limited due to the scarcity of available infected human tissue and methodological constraints in experimental cell models. *In vitro* cell models have demonstrated that DENV can infect skin cells [23], brain cells [24], dendritic cells [25], hepatic cells [26], among others. However, the presence of the virus in human tissues has been limited and does not necessarily indicate the lack of viral replication. One tissue that has attracted attention during DENV infection is the liver, where biochemical, histological, and morphological alterations have been observed. Most studies focus on DENV and YFV, while findings on ZIKV exposure remain limited and require further investigation.

Briefly, DENV infection has been associated with hepatocyte hyperplasia, hypertrophy, vacuolization, intracellular steatosis, hemorrhage, necrosis, inflammation, Kupffer cell apoptosis, and congestion of portal and centrilobular veins. Biochemical analyzes show insulin resistance and elevated alanine aminotransferase (ALT) and aspartate aminotransferase (AST) levels. Meanwhile, YFV infection is characterized by inflammation, apoptosis, and necrosis, along with elevated total bilirubin and ALT levels. ZIKV infection has been reported to cause hepatic inflammation with monocyte infiltration [27]. WNV, unlike the aforementioned *Orthoflavivirus*, can cause neuroinvasive diseases such as meningitis or encephalitis. In its most severe form, it can affect other organs, including the liver, with elevated AST and ALT levels [28].

## 4. General characteristics of DENV

Dengue is the most prevalent arboviral disease worldwide in terms of both incidence and mortality, and it is considered hyperendemic in tropical and subtropical regions, particularly in urban and semi-urban settings. DENV is fully adapted to humans and anthropophilic mosquitoes, particularly *Ae. aegypti* and *Ae. albopictus*. Four viral serotypes are recognized, all of which have also been identified in nonhuman primates [3].

Several studies have documented sylvatic cycles of DENV in Asia, with phylogenetic evidence supporting the origin of these serotypes from ancestral viruses circulating between primates and sylvatic mosquitoes. Currently, all four serotypes have evolved into genotypes highly adapted to human hosts [29].

### 4.1 Molecular structure of DENV

DENV virions are spherical, icosahedral, enveloped particles approximately 50 nm in diameter. They consist of an inner nucleocapsid, a sticky lipid envelope, and surface glycoproteins (membrane [M] and envelope [E] proteins). The E protein is essential for cell binding and membrane fusion, making it a primary target for neutralizing antibodies due to its immunogenicity [30].

The genome is a single-stranded, positive-sense RNA molecule with a 5′ cap and no 3′ poly-A tail. It encodes a single open reading frame (ORF) that produces a polyprotein, which is processed into three structural proteins (capsid [C], membrane [M], and envelope [E]) and seven nonstructural proteins (NS1, NS2A, NS2B, NS3, NS4A, NS4B, and NS5) [30]. Cleavage is mediated by viral proteases (NS2B/NS3) on the cytoplasmic side and host proteases (furin, signal peptidase) within the endoplasmic reticulum (ER) lumen [31].

Both immature and mature viral particles contain an outer membrane derived from the ER, incorporating E and M proteins to form the icosahedral glycoprotein surface. Viral maturation involves the pH-dependent rearrangement of E and M proteins in the trans-Golgi network (TGN), which determines viral infectivity. The E protein undergoes structural changes during this transition, shifting from an immature spiked form to a smooth mature form. Following maturation, the infectious peptide Pr—initially bound to the E protein—is cleaved by furin at neutral extracellular pH, producing fully infectious virions. NS2A and NS4B are involved in forming the viral replication complex [31, 32].

### 4.2 Viral replication

DENV replication occurs in the cytoplasm on host-derived ER membranes. The negative-sense RNA strand serves as a template for synthesizing positive-sense genomic RNA. Viral particles mature through the host's secretory and exocytic pathways, during which the prM protein is cleaved to form M protein, enabling the production of infectious virions [33].

Replication is facilitated by viral proteins (NS5, NS3, and others) and possibly host cell components. These proteins play multiple roles during infection and interact with both viral and host elements. In addition to NS1, NS2A, NS2B, NS4A, and NS4B, several proteins involved in the replication complex (RC) contribute to RNA amplification. Upon entry into the host cytoplasm, the viral genome immediately functions as mRNA [34].

Translation occurs on the rough ER (RER), inducing cellular membrane rearrangements that form virus-induced membrane structures—referred to as viral proteins (VPs)—that house the RC. Once enough viral proteins are produced, translation halts, and the RNA genome is redirected to these VPs for replication, although the exact mechanism of RNA trafficking remains unknown [34].

NS4A, NS4B, and possibly NS2A localize to the internal membranes of the VPs, whereas NS1 remains bound to the external ER-facing membrane. Viral polymerases NS5 and NS3 must be recruited into the VPs. It is likely that host factors and viral NS proteins aid in RNA transport and replication [32].

Within the VPs, the viral genome is initially translated as mRNA and subsequently used to synthesize new genomic RNA strands. NS5 binds the 5′ SLA (stem-loop A) region of the viral genome to initiate replication, which requires circularization of the RNA through complementary end pairing. Newly synthesized RNA is packaged by the C protein to form nucleocapsids, which acquire a lipid envelope by budding from the ER. During transit through the secretory pathway, the prM and E glycoproteins are structurally modified and cleaved, resulting in fully mature and infectious viral particles [32].

### 4.3 Pathogenesis

DENV transmission occurs primarily through the bite of infected female mosquitoes of the *Aedes* genus, as females are hematophagous, unlike the fruit-feeding males. Individuals of any age who are exposed to infected mosquitoes are susceptible to DENV infection. Although human-to-human transmission does not occur directly, cases of transmission through blood transfusions from infected donors to noninfected recipients have been reported [35].

The virulence of DENV is influenced by both viral and host factors, including preexisting immunity [36]. Certain DENV strains exhibit a greater capacity for replication in humans or vectors, enhancing transmission potential and increasing epidemic risk. Efficient replication in mosquitoes shortens the extrinsic incubation period, facilitating viral dissemination [32].

Disease severity in patients has been linked to viremia levels. Increased viral replication in human hosts may result from three key mechanisms, each of which enhances target cell infection, amplifies cytokine cascades, and increases pathogenicity [35–37]:

1. High replication rates: Pathogenic DENV strains replicate more rapidly in human cells, resulting in higher viral titers.

2. Immune evasion: Pathogenic strains can evade adaptive and cross-reactive host immune responses.

3. Antibody-dependent enhancement (ADE): This phenomenon involves non-neutralizing antibodies from a prior infection facilitating viral entry into host cells, thereby increasing viral replication. ADE is associated with severe clinical outcomes such as dengue hemorrhagic fever (DHF) and dengue shock syndrome (DSS), particularly in secondary infections with a different serotype and in cases of suboptimal immune responses.

### 4.4 Mechanisms of DENV damage to the liver

It is well established that DENV can replicate *in vitro* in hepatoma-derived cell lines such as HepG2 [38] and Huh-7 [39], as well as in human liver sinusoidal endothelial cells (HLSECs) [40]. These findings have prompted investigations into whether similar phenomena occur in experimental infection models. However, the development of an experimental model that faithfully reproduces the full clinical and pathological spectrum of human DENV infection remains a significant challenge.

Among the available models, DENV inoculation in BALB/c mice *via* intraperitoneal or intravenous routes has been utilized. Intraperitoneal inoculation has been associated with detectable viremia and focal hepatic lesions [41]. Intravenous administration, in contrast, results in more pronounced liver injury, including steatosis, hepatocellular swelling, and necrosis, primarily centered around the central vein and portal tracts [42]. These findings support the hypothesis that DENV is capable of hepatic replication *in vivo*.

To further evaluate this hypothesis, liver tissues from 13 pediatric patients who succumbed to dengue at Yangon Children's Hospital in Myanmar were analyzed [43]. Hepatomegaly of at least 2 cm was noted in all but one case, for which clinical data were unavailable. Histological sections and immunofluorescence staining revealed the presence of the viral genome from all four DENV serotypes. Additionally, both structural (E protein) and nonstructural viral proteins (NS1 and NS3) were detected. The findings indicated active infection in hepatocytes and Kupffer cells, while no evidence of infection was found in endothelial cells.

These data confirm that DENV can directly infect and replicate within hepatocytes and Kupffer cells, inducing cell death through apoptosis and, in some cases, necrosis. Viral infection is known to upregulate pro-inflammatory cytokines, leading to mitochondrial alterations in hepatocytes that trigger programmed cell death pathways and contribute to insulin resistance [27].

Metabolic reprogramming during DENV infection has also been documented in Huh-7 cells. The virus markedly inhibits glutamine oxidation and enhances glucose metabolism; however, glucose is redirected not for adenosine triphosphate (ATP) production but for endogenous lipid oxidation. As a result, fatty acids become the primary energy substrate in infected cells [44]. Lipid metabolism is significantly disrupted, leading to excessive lipid droplet accumulation in hepatocytes. This alteration supports viral replication by facilitating capsid protein segregation and genome encapsidation [27, 45]. In summary, DENV infection induces glycolytic modulation [46], mitochondrial dysfunction [47], activation of fatty acid synthesis [48], and lipid droplet accumulation [45].

Hepatic damage is further exacerbated by the deposition of complement components in hepatocytes, originating from both the classical and alternative pathways [43]. Additionally, DENV infection has been shown to increase oxidative stress in hepatic cells [49].

Hepatitis is a frequent complication of dengue and may be exacerbated by palliative treatment with paracetamol, which is recognized as the leading cause of drug-induced liver injury and acute liver failure in both the United States and Europe [50]. **Figure 2** illustrates some mechanisms of DENV damage to the liver.

### 4.5 Diagnosis of DENV infection

The diagnosis of DENV infection should be made primarily clinically, and management of the patient with suspected dengue should begin. Laboratory tests are performed for epidemiological surveillance purposes [51].

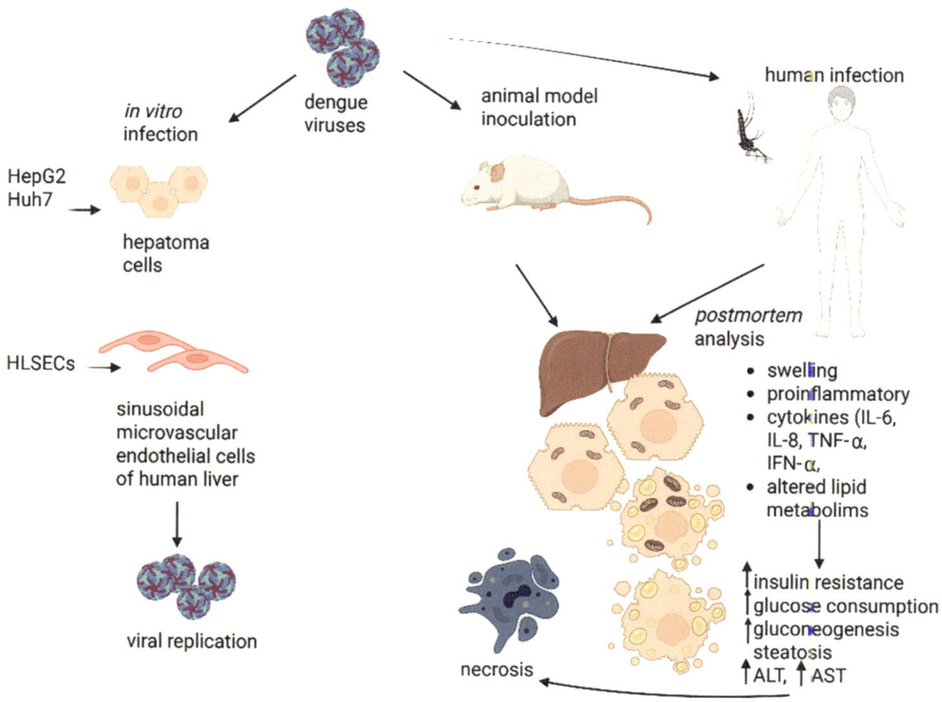


**Figure 2.**
*Mechanisms of DENV damage to the liver. The left side shows liver cells susceptible to infection* in vitro. *The right side illustrates the mouse model and postmortem findings in infected patients.*

#### *4.5.1 Clinical evaluation*

During clinical evaluation, it should be noted that some manifestations of dengue can be confused with other diseases caused by arboviruses such as ZIKV and Chikungunya virus; by other viruses such as rubella, measles, or influenza; by parasites such as *Plasmodium*, which causes malaria; and by bacteria such as *Salmonella enterica* serovar Typhi, which causes typhoid fever, among others. Therefore, a history should be conducted that includes possible contact with any of the aforementioned pathogens or any recent travel to areas where DENV is endemic. Additionally, the local epidemiological context should be explored. The physical examination should include body temperature (to rule out fever) and a search for skin rashes and minor bleeding. In the symptoms, one should ask if there is a headache, retroocular pain, myalgia, or arthralgia.

#### *4.5.2 Laboratory tests*

They can be performed on blood samples, autopsy tissue, and mosquitoes (these last two are regularly used for epidemiological control). A blood sample taken within the first 5 days of the onset of symptoms is called the acute phase of the disease. A blood sample taken 2–3 weeks after the onset of symptoms is considered a convalescent phase sample.

##### *4.5.2.1 Detection of genetic material*

The virus's genetic material (viral RNA) is detected in the early stages of the disease by the RT-PCR (reverse transcriptase-polymerase chain reaction) test. It is performed on acute-phase blood samples, autopsy tissue, and for epidemiological monitoring of mosquitoes. RT-PCR is a rapid, sensitive (nearly 100% detection), simple, and reproducible method with appropriate controls.

##### *4.5.2.2 NS1 viral protein detection*

This test detects the presence of viral antigen in the blood of an infected patient during the first 7 days of infection. It is performed using an enzyme-linked immunosorbent assay (ELISA).

##### *4.5.2.3 Viral antigen detection by immunohistochemistry*

This technique allows the detection of viral antigens in a wide variety of tissues, usually from deceased patients. Therefore, it is used to confirm DENV infection.

##### *4.5.2.4 Antibody determination*

This involves detecting IgM and IgG antibodies against the DENV in the serum of an acute-phase blood sample. Antibodies have even been detected up to 12 weeks after the onset of infection. A positive test result indicates recent or current infection. The recommended technique is the capture ELISA or other variants thereof.

#### *4.5.2.5 Viral isolation*

For this test, biological samples that may contain the virus are collected, such as blood, serum, cerebrospinal fluid, or autopsy tissue. Generally, samples must be taken within the first 5 days of presenting symptoms of infection. The samples are processed and placed in contact with virus-permissive cells such as monkey kidney cells, mosquito cells, and human fibroblasts, among others. The cells are incubated under conditions of temperature, $CO_2$, and humidity appropriate for virus replication. The cells are monitored to detect changes due to viral infection, and the isolated virus is subsequently characterized. This test can take more than a week, so it is recommended for epidemiological monitoring.

The World Health Organization recommends that laboratory resources be used to confirm all cases of dengue with warning signs and severe dengue, as well as all deaths [51].

### 4.6 Treatment of DENV infection

There is no specific antiviral against DENV. Once DENV infection is suspected, management focuses primarily on symptomatic treatment. The goal is to relieve pain, control fever, maintain adequate hydration, and recommend rest to prevent complications. Warning signs (severe and persistent abdominal pain, vomiting, edema, mucosal bleeding, altered consciousness, irritability, drowsiness, hepatomegaly, and progressively increased hematocrit) should be monitored to prevent progression to severe forms of dengue.

The most used drug to relieve pain (headache, arthralgia, and myalgia) and control fever is paracetamol. However, care must be taken not to exceed the recommended dose, as this can cause severe liver damage. Additionally, liver function should be monitored, considering a history of alcoholism and current DENV infection, which can alter liver metabolism.

It is important to avoid self-medication, the use of scientifically unvalidated home remedies, or even the use of other analgesics such as aspirin or ibuprofen, as these can increase the risk of hemorrhagic complications associated with dengue.

## 5. Factors involved in liver injury

It is important to highlight that the severity of hepatic damage in DENV infection can vary and is influenced by multiple factors, including host-related, vector-related, and environmental determinants. In animal models, outcomes depend largely on the type of model used—such as BALB/c, AG129, SCID, transgenic, humanized mice, or even nonhuman primates—as well as on the route of viral administration (intraperitoneal, intracerebral, intravenous, etc.) and age. Viral factors also play a role, including the serotype involved, the viral load inoculated, and the origin of the virus (i.e., patient-derived isolates versus laboratory strains). Notably, in an experimental model using rhesus macaques, all four DENV serotypes were found to affect liver transaminases [52].

In humans, the variability in liver injury has been associated with environmental and viral factors across all DENV serotypes [43]. These associations have been more extensively studied in endemic areas, where epidemiological surveillance facilitates assessment. In such settings, it is critical to evaluate prior exposure to other serotypes or even to different flaviviruses that may elicit cross-reactive immune responses, to rule out coinfections with other viruses, to analyze the individual host immune response, and to consider comorbid conditions, the use of hepatotoxic medications, the viral load transmitted by the mosquito, and even prior alcohol consumption that may have caused preexisting hepatic damage.

Host-related factors must be carefully evaluated during physical examination and clinical follow-up, especially for the presence of warning signs of severe dengue. These include intense abdominal pain, persistent vomiting, rapid breathing, bleeding from the gums or nose (mucosal bleeding), fatigue, restlessness, hematemesis or melena, excessive thirst, pale and cold skin, and generalized weakness [53].

Recently, a study aimed at predicting dengue severity analyzed 500 patients and identified several clinical predictors of severe disease. These included older age, elevated white blood cell count, increased hematocrit, prolonged prothrombin time, thrombocytopenia, hypotension, presence of perioral edema, third-space fluid loss, hepatomegaly, and involvement of other organs [54].

The role of the liver in predicting dengue severity is undeniable. However, it is equally important to assess prior conditions that could have compromised liver function, such as alcohol consumption, which not only predisposes the liver to injury but also shares several pathogenic mechanisms with DENV. The following section outlines the mechanisms by which ethanol induces liver damage.

### 5.1 Liver damage due to ethanol

Upon ingestion, ethanol (also known as ethyl alcohol) triggers a cascade of biochemical reactions that affect multiple organ systems, ultimately leading to liver diseases such as alcoholic hepatitis and cirrhosis. Although considerable knowledge exists regarding the pathophysiological mechanisms induced by ethanol in the human body, no single pathway can fully explain the wide range of adverse effects exerted by ethanol on various organs and tissues.

One key factor extensively studied and central to many ethanol-induced effects is the excessive generation of reactive molecules known as free radicals. These molecules contribute to a condition termed oxidative stress, which disrupts normal biochemical processes within the cell and can ultimately lead to programmed cell death, or apoptosis [6, 10].

The mitochondrial respiratory chain, one of the main physiological sources of superoxide anion production, is particularly susceptible to ethanol exposure. Ethanol enhances the rate of reactive oxygen species (ROS) generation in mitochondria (see **Figure 3**). Mitochondria contain the enzyme superoxide dismutase (SOD), which converts superoxide into hydrogen peroxide. This hydrogen peroxide is subsequently detoxified by glutathione peroxidase. However, in the presence of iron, part of the hydrogen peroxide escapes degradation and forms highly reactive oxygen species that can induce both structural and functional mitochondrial alterations.

Experimental administration of xenobiotics has shown increased superoxide production in liver mitochondria and isolated hepatocytes. The accumulation of superoxide may reduce the nicotinamide adenine dinucleotide ($NAD^+$)/NADH ratio, disrupting the mitochondrial redox balance and increasing electron flow along the respiratory chain. This, in turn, leads to the formation of hydroxyl radicals, contributing further to mitochondrial lipid peroxidation. The shift in redox

*DOI: http://dx.doi.org/10.5772/intechopen.1011926*

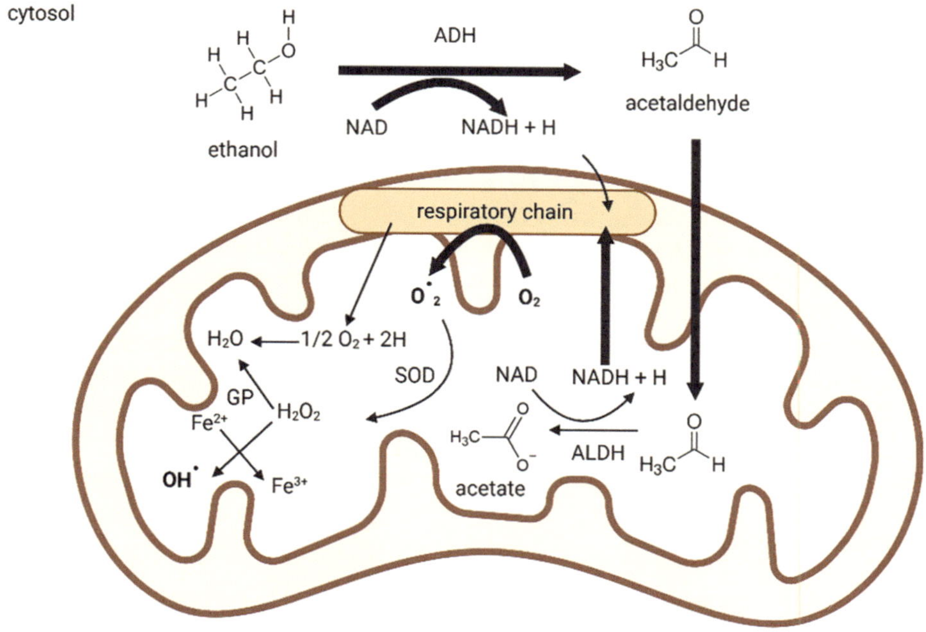


**Figure 3.**
*Effect of ethanol on free radical production in mitochondria.*

potential, combined with damage to inner membrane lipids, impairs mitochondrial energy synthesis by decreasing the membrane potential. Chronic ethanol exposure has been shown to induce morphological and structural changes in hepatic mitochondria, representing some of the earliest manifestations of alcohol-induced liver damage [6, 10].

When ethanol is ingested, it is converted into acetaldehyde in the cytosol of hepatocytes *via* the ADH, with the involvement of $NAD^+$. Acetaldehyde has been shown to increase the production of alkanes and alter the cytosolic redox potential by shifting the $NAD^+/NADH$ ratio, similar to its effects in mitochondria. This redox imbalance promotes lipid peroxidation and exacerbates cellular damage. Additionally, the ALDH plays a pivotal role in altering the redox potential and contributing to oxidative stress in the cytosol. The Km of ALDH for acetaldehyde—the primary product of ethanol oxidation by ADH—is approximately 1 mM, significantly lower than that of xanthine oxidase. As a result, ALDH more readily converts acetaldehyde into acetate, using $NAD^+$ in its catalytic process and further disrupting the redox balance [6, 10].

Based on the above, ethanol metabolism in the liver represents the primary route for the elimination of this xenobiotic substance. This metabolic pathway is influenced by various physiological (e.g., age, sex) and pathophysiological (e.g., cirrhosis, gastritis) conditions, as well as by external factors, such as the use of medications in chronic or infectious diseases. Ethanol metabolism disrupts several critical pathways essential for hepatic cytophysiology, contributing to cellular imbalance and the pathogenesis of alcohol-related liver disease. Although the metabolism of ethanol occurs in seemingly straightforward steps, it intersects with major metabolic pathways—carbohydrate, lipid, and protein metabolism. Consequently, alcohol consumption may alter the natural history of a disease when a virus that affects liver function is present, such as DENV, whether infection leads to clinical illness [55].

## 6. Conclusion

DENV infection causes multiple forms of liver damage, ranging from asymptomatic biochemical alterations and inflammation to histological changes and acute liver failure. The severity of hepatic injury can vary and depends on multiple factors, one of the most common yet underexplored being alcohol consumption, whether chronic or occasional. The evidence indicates that both DENV infection and alcohol share common pathways of liver injury, including increased oxidative stress and the induction of inflammatory cytokine production, which lead to cell death *via* apoptosis and necrosis. These processes are ultimately reflected in elevated hepatic transaminases, steatosis due to the activation of gluconeogenesis, increased ATP demand, and consequent activation of glycolytic pathways. Therefore, from the clinical diagnosis of DENV infection, a history of alcohol consumption should be questioned, and the metabolic status of the liver should be monitored during treatment.

## Acknowledgements

We gratefully acknowledge CONAHCYT, Instituto Politécnico Nacional, and SIP/IPN grants: 20250361 and 20254045.

## Conflict of interest

The authors declare no conflict of interest.

## Nomenclature

| | |
|---|---|
| ADE: | antibody-dependent enhancement |
| ADH: | alcohol dehydrogenase |
| ALDH: | aldehyde dehydrogenase |
| ALT: | alanine aminotransferase |
| AST: | aspartate aminotransferase |
| ATP: | adenosine triphosphate |
| C: | capsid protein |
| DENV: | dengue virus |
| DHF: | dengue hemorrhagic fever |
| DSS: | dengue shock syndrome |
| DTV: | deer tick virus |
| E: | envelope protein |
| ER: | endoplasmic reticulum |
| GHRF: | growth hormone–releasing factor |
| HDL: | high-density lipoproteins |
| HLSECs: | human liver sinusoidal endothelial cells |
| JEV: | Japanese encephalitis virus |
| LDL: | low-density lipoproteins |
| M: | membrane protein |
| MEOS: | microsomal ethanol oxidizing system |
| $NAD^+$: | nicotinamide adenine dinucleotide |

| | |
|---|---|
| NADH: | reduced form of nicotinamide adenine dinucleotide |
| NK: | natural killer cell |
| NS: | nonstructural protein |
| ORF: | open reading frame |
| POWV: | powassan virus |
| prM: | premembrane protein |
| RC: | replication complex |
| RER: | rough endoplasmic reticulum |
| ROS: | reactive oxygen species |
| SLA: | stem-loop A |
| SOD: | superoxide dismutase |
| T3: | triiodothyronine |
| T4: | thyroxine |
| TBEV: | tick-borne encephalitis virus |
| TGN: | trans-Golgi network |
| VLDL: | very-low-density lipoproteins |
| VPs: | viral proteins |
| WNV: | West Nile virus |
| YFV: | yellow fever virus |
| ZIKV: | Zika virus |

## Author details

Jazmín García-Machorro[1], Gabriela Mellado-Sánchez[2,3]
and José Antonio Morales-González[1*]

1 Laboratorio de Medicina de la Conservación, Escuela Superior de Medicina, Instituto Politécnico Nacional, Mexico City, Mexico

2 Unidad de Desarrollo e Investigación en Bioterapéuticos (UDIBI), Escuela Nacional de Ciencias Biológicas, Instituto Politécnico Nacional, Mexico City, Mexico

3 Laboratorio Nacional para Servicios Especializados de Investigación, Desarrollo e Innovación (I+D+i) para Farmoquímicos y Biotecnológicos, LANSEIDI-FarBiotec-CONAHCYT, Mexico City, Mexico

*Address all correspondence to: jmorales101@yahoo.com.mx

## References

[1] Mifsud JCO, Lytras S, Oliver MR, Toon K, Costa VA, Holmes EC, et al. Mapping glycoprotein structure reveals Flaviviridae evolutionary history. Nature. 2024;**633**(8030):695-703

[2] Postler TS, Beer M, Blitvich BJ, Bukh J, de Lamballerie X, Drexler JF, et al. Renaming of the genus Flavivirus to Orthoflavivirus and extension of binomial species names within the family Flaviviridae. Archives of Virology. 2023;**168**(9):224

[3] de Almeida PR, Weber MN, Sonne L, Spilki FR. Aedes-borne Orthoflavivirus infections in neotropical primates – Ecology, susceptibility, and pathogenesis. Experimental Biology and Medicine. 2023;**248**(22):2030-2038

[4] Mondet B, da Rosa AP, Vasconcelos PF. The risk of urban yellow fever outbreaks in Brazil by dengue vectors. *Aedes aegypti* and *Aedes albopictus*. Bulletin de la Societe de Pathologie Exotique. 1996;**89**(2):107-113; discussion 114

[5] Blahove MR, Carter JR. Flavivirus persistence in wildlife populations. Viruses. 2021;**13**(10):2099

[6] Morales González JA. Alcohol, alcoholismo y cirrosis. Un Enfoque Multidisciplinario. Vol. 240. Pachuca: Universidad Autónoma del Estado de Hidalgo; 2007

[7] Ahmadi M, Shirafkan H, Mozaffarpur SA, Rezghi M. Impact of jujube fruit on serum lipid profile, glycemic index, and liver function: A systematic review and meta-analysis. Nutrition & Diabetes. 2025;**15**(1):22

[8] Ackermann MR. Inflammation and healing. In: Zachary JF, editor. Pathologic Basis of Veterinary Disease. 6th ed. St Louis, MO: Elsevier; 2017. pp. 73-131.e2

[9] Ramírez-Farías C, Madrigal-Santillán E, Gutiérrez-Salinas J, Rodríguez-Sánchez N, Martínez-Cruz M, Valle-Jones I, et al. Protective effect of some vitamins against the toxic action of ethanol on liver regeneration induced by partial hepatectomy in rats. World Journal of Gastroenterology: WJG. 2008;**14**(6):899

[10] Piña-Garza E, Gutiérrez-Salinas J, Morales-González JA, Zentella de Piña M. Es tóxico el alcohol? In: Temas Bioquímicos de vanguardia. México: Facultad de Medicina, Universidad Nacional Autónoma de México; 2003. pp. 121-146

[11] Zakhari S. Overview: How is alcohol metabolized by the body? Alcohol Research & Health. 2006;**29**(4):245-254

[12] Morales-González JA, de Lourdes Sernas-Morales M, Morales-González Á, González-López LL, Madrigal-Santillán EO, Vargas-Mendoza N, et al. Morphological and biochemical effects of weekend alcohol consumption in rats: Role of concentration and gender. World Journal of Hepatology. 2018;**10**(2):297

[13] Contreras-Zentella ML, Villalobos-García D, Hernández-Muñoz R. Ethanol metabolism in the liver, the induction of oxidant stress, and the antioxidant defense system. Antioxidants. 2022;**11**(7):1258

[14] Wisse E, Luo D, Vermijlen D, Kanellopoulou C, de Zanger R, Braet F. On the function of pit cells, the liver-specific natural killer cells. Seminars in Liver Disease. 1997;**17**(04):265-286

[15] Parlar YE, Ayar SN, Cagdas D, Balaban YH. Liver immunity, autoimmunity, and inborn errors of immunity. World Journal of Hepatology. 2023;**15**(1):52-67

[16] Wang M, Feng Z. Mechanisms of hepatocellular injury in hepatitis A. Viruses. 2021;**13**(5):861

[17] Suhail M. Potential mechanisms of hepatitis B virus induced liver injury. World Journal of Gastroenterology. 2014;**20**(35):12462

[18] Khatun M, Ray RB. Mechanisms underlying hepatitis C virus-associated hepatic fibrosis. Cells. 2019;**8**(10):1249

[19] Diaz G, Engle RE, Tice A, Melis M, Montenegro S, Rodriguez-Canales J, et al. Molecular signature and mechanisms of hepatitis D virus–associated hepatocellular carcinoma. Molecular Cancer Research. 2018;**16**(9):1406-1419

[20] Irshad M, Sharma Y, Dhar I, Singh J, Joshi YK. Transfusion-transmitted virus in association with hepatitis A-E viral infections in various forms of liver diseases in India. World Journal of Gastroenterology. 2006;**12**(15):2432-2436

[21] Nanaware N, Banerjee A, Mullick Bagchi S, Bagchi P, Mukherjee A. Dengue virus infection: A tale of viral exploitations and host responses. Viruses. 2021;**13**(10):1967

[22] Mangat R, Louie T. Arbovirus encephalitides. In: StatPearls [Internet]. Treasure Island (FL): StatPearls Publishing; 2023

[23] Bustos-Arriaga J, García-Machorro J, León-Juárez M, García-Cordero J, Santos-Argumedo L, Flores-Romo L, et al. Activation of the innate immune response against DENV in normal non-transformed human fibroblasts. PLoS Neglected Tropical Diseases. 2011;**5**(12):e1420

[24] Ho MR, Tsai TT, Chen CL, Jhan MK, Tsai CC, Lee YC, et al. Blockade of dengue virus infection and viral cytotoxicity in neuronal cells in vitro and in vivo by targeting endocytic pathways. Scientific Reports. 2017;**7**(1):6910

[25] Helgers LC, Keijzer NCH, van Hamme JL, Sprokholt JK, Geijtenbeek TBH. Dengue virus infects human skin Langerhans cells through Langerin for dissemination to dendritic cells. Journal of Investigative Dermatology. 2024;**144**(5):1099-1111.e3

[26] Ferreira JGG, Gava SG, Oliveira ES, Batista ICA, Fernandes G, d R, Mourão MM, et al. Gene expression signatures in AML-12 hepatocyte cells upon dengue virus infection and acetaminophen treatment. Viruses. 2020;**12**(11):1284

[27] Pinheiro BSS, Rodrigues JG, Dias FCR, de Oliveira Gomes A, de Lucca Moreira Gomes M. Hepatic damage caused by flaviviruses: A systematic review. Life Sciences. 2023;**331**:122074

[28] Bruno L, Nappo MA, Frontoso R, Perrotta MG, Di Lecce R, Guarnieri C, et al. West Nile virus (WNV): One-health and eco-health global risks. Veterinary Sciences. 2025;**12**(3):288

[29] Nyathi S, Rezende IM, Walter KS, Thongsripong P, Mutuku F, Ndenga B, et al. Geographic origin and evolution of dengue virus serotypes 1 and 3 circulating in Africa. Virus Evolution. 2025;**11**(1):veae116

[30] Li Q, Kang C. Structures and dynamics of dengue virus nonstructural membrane proteins. Membranes (Basel). 2022;**12**(2):231

[31] Braun NJ, Quek JP, Huber S, Kouretova J, Rogge D, Lang-Henkel H, et al. Structure-based macrocyclization of substrate analogue NS2B-NS3 protease inhibitors of Zika, West Nile and dengue viruses. ChemMedChem. 2020;**15**(15):1439-1452

[32] Pourzangiabadi M, Najafi H, Fallah A, Goudarzi A, Pouladi I. Dengue virus: Etiology, epidemiology, pathobiology, and developments in diagnosis and control – A comprehensive review. Infection, Genetics and Evolution. 2025;**127**:105710

[33] Mansfield KL, Horton DL, Johnson N, Li L, Barrett ADT, Smith DJ, et al. Flavivirus-induced antibody cross-reactivity. The Journal of General Virology. 2011;**92**(Pt 12):2821-2829

[34] Ferrari M, Zevini A, Palermo E, Muscolini M, Alexandridi M, Etna MP, et al. Dengue Virus Targets Nrf2 for NS2B3-Mediated Degradation Leading to Enhanced Oxidative Stress and Viral Replication. Journal of Virology. 23 Nov 2020;**94**(24):e01551-20

[35] Islam MT, Quispe C, Herrera-Bravo J, Sarkar C, Sharma R, Garg N, et al. Production, Transmission, Pathogenesis, and Control of Dengue Virus: A Literature-Based Undivided Perspective. BioMed Research International. 15 Dec 2021;**2021**:4224816

[36] Bhatt P, Sabeena SP, Varma M, Arunkumar G. Current understanding of the pathogenesis of dengue virus infection. Current Microbiology. 2021;**78**(1):17-32

[37] Kok BH, Lim HT, Lim CP, Lai NS, Leow CY, Leow CH. Dengue virus infection – A review of pathogenesis, vaccines, diagnosis and therapy. Virus Research. 2023;**324**:199018

[38] Thepparit C, Phoolcharoen W, Suksanpaisan L, Smith DR. Internalization and propagation of the dengue virus in human hepatoma (HepG2) cells. Intervirology. 2004;**47**(2):78-86

[39] Pando-Robles V, Oses-Prieto JA, Rodríguez-Gandarilla M, Meneses-Romero E, Burlingame AL, Batista CVF. Quantitative proteomic analysis of Huh-7 cells infected with dengue virus by label-free LC–MS. Journal of Proteomics. 2014;**111**:16-29

[40] Puerta-Guardo H, Glasner DR, Espinosa DA, Biering SB, Patana M, Ratnasiri K, et al. Flavivirus NS1 triggers tissue-specific vascular endothelial dysfunction reflecting disease tropism. Cell Reports. 2019;**26**(6):1598-1613.e8

[41] Yauch L, Shresta S. Mouse models of dengue virus infection and disease. Antiviral Research. 2008;**80**(2):87-93

[42] Paes MV, Lenzi HL, Nogueira ACM, Nuovo GJ, Pinhão ÂT, Mota EM, et al. Hepatic damage associated with dengue-2 virus replication in liver cells of BALB/c mice. Laboratory Investigation. 2009;**89**(10):1140-1151

[43] Aye KS, Charngkaew K, Win N, Wai KZ, Moe K, Punyadee N, et al. Pathologic highlights of dengue hemorrhagic fever in 13 autopsy cases from Myanmar. Human Pathology. 2014;**45**(6):1221-1233

[44] Fernandes-Siqueira LO, Zeidler JD, Sousa BG, Ferreira T, Da Poian AT. Anaplerotic Role of Glucose in the Oxidation of Endogenous Fatty Acids during Dengue Virus Infection. mSphere. 2018;**3**(1):e00458-17

[45] Samsa MM, Mondotte JA, Iglesias NG, Assunção-Miranda I, Barbosa-Lima G, Da Poian AT, et al. Dengue virus capsid

protein usurps lipid droplets for viral particle formation. PLoS Pathogens. 2009;**5**(10):e1000632

[46] Fontaine KA, Sanchez EL, Camarda R, Lagunoff M. Dengue virus induces and requires glycolysis for optimal replication. Journal of Virology. 2015;**89**(4):2358-2366

[47] El-Bacha T, Midlej V, Pereira da Silva AP, Silva da Costa L, Benchimol M, Galina A, et al. Mitochondrial and bioenergetic dysfunction in human hepatic cells infected with dengue 2 virus. Biochimica et Biophysica Acta (BBA) - Molecular Basis of Disease. 2007;**1772**(10):1158-1166

[48] Heaton NS, Perera R, Berger KL, Khadka S, LaCount DJ, Kuhn RJ, et al. Dengue virus nonstructural protein 3 redistributes fatty acid synthase to sites of viral replication and increases cellular fatty acid synthesis. Proceedings of the National Academy of Sciences. 2010;**107**(40):17345-17350

[49] Wang J, Chen Y, Gao N, Wang Y, Tian Y, Wu J, et al. Inhibitory effect of glutathione on oxidative liver injury induced by dengue virus serotype 2 infections in mice. PLoS One. 2013;**8**(1):e55407

[50] Kellstein D, Fernandes L. Symptomatic treatment of dengue: Should the NSAID contraindication be reconsidered? Postgraduate Medicine. 2019;**131**(2):109-116

[51] Organización Panamericana de la Salud. Hoja informativa sobre el dengue [Internet]. Available from: https://www.paho.org/es/temas/dengue [Accessed: July 04, 2025]

[52] Hickey AC, Koster JA, Thalmann CM, Hardcastle K, Tio PH, Cardosa MJ, et al. Serotype-specific host responses in rhesus macaques after primary dengue challenge. The American Society of Tropical Medicine and Hygiene. 2013;**89**(6):1043-1057

[53] World Health Organization. Dengue and Severe Dengue. 2024. [Internet]. Available from: https://www.who.int/news-room/fact-sheets/detail/dengue-and-severe-dengue [Accessed: July 04, 2025]

[54] Bhandari S, Rankawat G, Goyal B, Lohmror A, Gupta V, Singh A. Establishment of SMS dengue severity score. Journal of the Association of Physicians of India. 2022;**70**(9):19-22

[55] Rodríguez-Negrete EV, García-Machorro J, Madrigal-Santillán EO, Morales-González Á, Morales-González JA. Effect of weekend alcohol consumption on hepatic antioxidant enzyme activity: Role of concentration and gender. Gastroenterology Insights. 2025;**16**(2):13

Chapter 8

# Update on Vertical Transmission and Coinfection of Zika and Dengue Viruses in *Aedes aegypti* and Research on Botanical Extracts as Alternative Vector Control Strategies

*Josselin Carolina Corzo-Gómez, Mayra Montecillo-Aguado, Esther Candelaria Martínez-Molina, Sury Antonio López-Cancino, Victor Javier Cruz-Holguin, Giovani Visoso-Carvajal, Alejandra García-Aguilar, María Fernanda García-Aguilar and Sandra Paola Martínez-Frías*

## Abstract

*Aedes aegypti* has an effective vector capacity for the transmission of dengue and Zika viruses in humans as well as in primates; due to its intradomiciliary and hematophagous behavior with multiple human hosts, it increases the probability of acquiring and maintaining both viruses in nature, benefiting their dissemination, in addition to facilitating vertical transmission to the offspring, this becomes relevant due to its marked preference for human blood, essential for the development of its eggs, since during each oviposition, it lays an average of between 100 and 150 eggs, which could affect the population density of probably infected mosquitoes, increasing the risk of transmission to humans, situations that favor both a persistent circulation of these viruses, as well as a wide geographical distribution. Therefore, it is important to update knowledge about the vector characteristics that can favor the transmission of these two pathogens, highlighting the need to further develop integrated vector control strategies, addressing them with an ecological approach, with the aim of preventing the vector from contacting its host as quickly as possible, without affecting human health or the environment. In this regard, the importance of current research on plant extracts and essential oils that can provide protection against mosquito bites is highlighted. These can offer a sustainable and natural



IntechOpen

alternative to synthetic pesticides, as they not only combat mosquitoes at their different stages of development but can also help minimize environmental damage, having the potential to be used as an ecological approach to *Ae. aegypti* control.

**Keywords:** *Aedes aegypti*, Zika virus, dengue virus, vertical transmission, coinfection, plant extracts, integrated control, ecological approach

## 1. Introduction

Dengue virus (DENV) and Zika virus (ZIKV) are important *Orthoflaviviruses* transmitted mainly by female *Ae. aegypti* mosquitoes and, when disseminated by this particular category of arthropods, they are likewise referred to as arboviruses, which in recent years have proven to be serious public health problems, especially in tropical regions. This is of greater importance because both viruses share a vector, which can complicate transmission due to their cocirculation and coexistence among the population. Considering that infection with both viruses in humans can cause similar clinical symptoms, and although no alterations or complications in the clinical presentation have been observed in coinfections in most infected patients, a smaller proportion of the affected population has presented clinical complications. Therefore, it is important to understand the epidemiology, behavior, and characteristics of the vector. This knowledge is essential for addressing and managing alternative control strategies for the different life stages of the vector, making outbreak management more effective. In this regard, it is important to consider that there is currently a deficiency or absence of an effective vaccines to prevent these diseases, and that both pathologies lack specific antiviral treatments, so vector control remains an important strategy. On the other hand, biotic and abiotic factors play an important role in the complex interaction between the vector and these viruses, which determines vector competence. Together, they define the mosquito's vectorial capacity to transmit these pathogens to the human population due to the vector's bionomics, which includes age, lifespan, number of gonotrophic cycles, hematophagous habits, as well as the density and population structure of mosquitoes, in addition to other factors such as temperature, humidity, type of population (rural or urban), or the characteristics of the viruses themselves, which have been considered determining factors in the transmission and dispersion process of these arboviruses.

Historically, DENV epidemics, and currently those of ZIKV, have been controlled through the application of insecticides, which has led to the gradual establishment of resistance against numerous conventional chemical insecticides. This has also led to the observation of toxicological repercussions on the food chain and environmental quality, which impact human and animal health. Therefore, the use of plant-derived extracts and essential oils containing bioactive compounds presents a viable alternative methodology for the control of mosquito vector populations [1], which also justifies the use of alternative products that are safer for the environment and pose a lower risk to human health.

## 2. Epidemiology and transmission cycles of DENV and ZIKV

### 2.1 Epidemiology of dengue and Zika disease

Dengue fever is one of the most common mosquito-borne viral infections in tropical and subtropical areas worldwide. Among the most reported endemic areas are Southeast Asia, the Western Pacific, Latin America, the Caribbean, and parts of Africa. The World Health Organization (WHO) estimates that around 50% of the world's population lives in areas at risk of dengue fever, and an estimated 100 to 400 million dengue infections occur annually [2]. Around 500,000 cases of severe dengue require hospitalization, and approximately 20,000 people die from complications. The most affected areas include Latin America (Brazil, Mexico, Colombia, Honduras), Asia (India, Thailand, Indonesia, the Philippines), and Africa (Sudan, Angola, Kenya) [3]. Dengue infection is caused by a virus of the *Orthoflavivirus* genus, which includes four distinct serotypes: DENV1, DENV2, DENV3, and DENV4. These viruses are closely related and are primarily transmitted by the bite of an infected *Ae. aegypti* mosquitoes. Each serotype can cause the disease, but infection with one serotype does not guarantee immunity against the others, which can result in more severe secondary infections [4, 5]. This virus is an enveloped virus with a spherical morphology and a diameter of approximately 50 nm. Its viral genome consists of a single strand of positive-sense RNA, 11 kb in size, and is composed of positive-strand single-stranded RNA, which encodes a single polyprotein of 3391 amino acids that is processed by cellular proteases and by the viral protease NS2B-3 to give rise to three structural proteins: capsid (C), membrane precursor (prM) and envelope (E), which make up the structure of the virion. Meanwhile, the seven non-structural proteins: NS1, NS2A, NS2B, NS3, NS4A, NS4B, and NS5 are involved in the process of viral replication [6, 7].

In the context of ZIKV, despite its initial identification occurring in Uganda in the year 1947, its proliferation on a global scale commenced in 2007, with significant outbreaks in Latin America and the Caribbean. Among the most reported endemic areas are Latin America (Brazil, Colombia, Mexico, Venezuela, El Salvador), the Caribbean (Puerto Rico, Dominican Republic, Haiti), Asia (Philippines, Thailand, Malaysia), and Africa (Uganda, Cape Verde, Angola). It is estimated that during a massive epidemic in Latin America in 2015–2016, more than 500,000 cases were reported. In 2016, the WHO declared a public health emergency of international concern due to the development of congenital Zika syndrome, frequently including microcephaly, and neurological disorders, as damage to the developing fetal brain, specifically affecting the growth of neuronal cells. Because of this, and although cases from the 2015–2016 outbreak decreased considerably, the virus continues to circulate in tropical areas, with sporadic outbreaks still being of significant concern [8–10]. Regarding ZIKV, it has been classified into two main lineages: the Asian and African lineages. The ZIKV circulating in the Americas is phylogenetically closer to the Asian lineage [11, 12]. The primary mode of transmission for this virus is through the bite of infected *Aedes* mosquitoes. Zika virus is a positive-sense, single-stranded RNA virus. The genome of this virus contains 10,794 nucleotides encoding 3419 amino acids. It is composed of two non-coding regions flanking an open reading frame encoding a polyprotein that gives rise to a capsid protein, a membrane precursor protein, an envelope protein, and seven non-structural proteins [9].

### 2.2 Transmission cycles of DENV and ZIKV

Two distinct transmission cycles have been identified and documented for both viral agents: the urban cycle (the most common cycle), where humans act as the primary hosts; transmission occurs primarily in densely populated areas with a high presence of *Ae. aegypti*. The sylvatic or zoonotic cycle has been identified in regions of

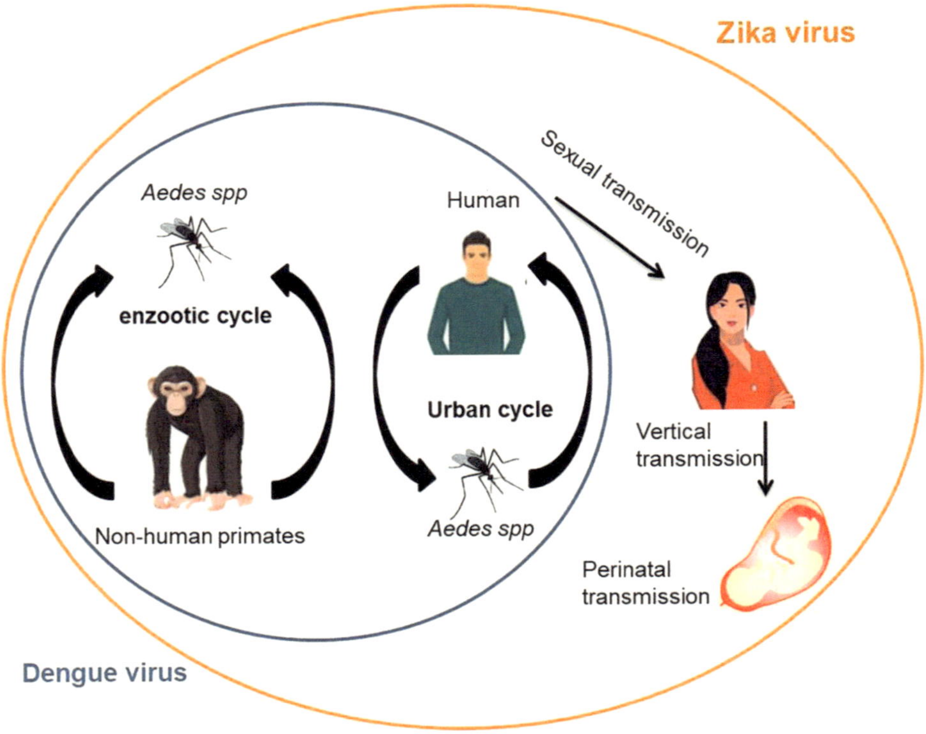


**Figure 1.**
*Main transmission routes for DENV and ZIKV include the vector-borne route and secondary routes such as vertical transmission, transmission of the virus from infected mother to fetus, and, in the case of ZIKV, sexual transmission. Prepared by the authors.*

Africa and Southeast Asia (**Figure 1**). Within this cycle, non-human primates act as natural reservoirs for the viruses, with sylvatic mosquitoes facilitating their transmission to other primate hosts [13, 14].

The transmission cycle of DENV and ZIKV occurs through the bite of previously infected female mosquitoes on healthy hosts. The urban cycle involves mosquito-to-human-to-mosquito transmission, and various phases of this cycle are recognized. In the first phase, the vector infection is described, which occurs when a mosquito bites an infected person during the viremia phase; subsequently, the virus enters the mosquito's intestine and begins to replicate, giving rise to the incubation phase, which in the particular case of ZIKV requires an extrinsic incubation period of 4–10 days before the mosquito can transmit it, in the case of DENV the incubation period within the mosquito varies from 6 to 15 days, depending on the ambient temperature and other factors. Subsequently, the virus migrates to the salivary glands until it encounters the human host, and finally, the infected mosquito bites a healthy person, introducing the virus into the bloodstream. The primary targets for both viruses are dermal fibroblasts, keratinocytes, monocytes, and dendritic cells. However, particularly for ZIKV, it can also infect neuronal precursor cells. In infected cells, the viruses bind to cellular receptors, which are diverse, and some depend on the target cell type, such as DC-SIGN, heparan sulfate, or mannose receptors [15]. Following a successful infection, the processes of replication and the synthesis of novel viral particles commence. The duration of viremia has been established to persist for an approximate period of 4–7 days [16, 17].

Consequently, the predominant mechanism for the transmission of DENV and ZIKV is through the bite of their shared vector, *Ae. aegypti*, however, in the case of DENV, vertical transmission of the virus has also been reported during pregnancy and through blood transfusions. In the case of ZIKV, alternative modes of transmission have also been documented, specifically through direct pathways, namely, from individual to individual, which transpires perinatally, *via* sexual contact, and through breastfeeding or blood transfusions, thereby significantly augmenting the incidence of both mild and severe cases, particularly among neonates [18, 19].

## 3. Clinical presentation of infection and coinfection in humans and in *in vitro* and *in vivo* models

### 3.1 Pathogenesis of DENV and ZIKV infections

Arboviruses within the genus *Orthoflavivirus*, like DENV and ZIKV, represent a major public health challenge in Latin America and worldwide due to their negative impact on the health of the population. Infectious diseases caused by these viruses present a wide variety of clinical symptoms, producing a similar clinical picture in the first days of incubation of the infection. Despite the significant challenge of differential diagnosis, it is critical for healthcare providers to identify the symptoms generated by DENV and ZIKV infections, enabling timely diagnosis and appropriate care and treatment. Addressing the interaction or coinfection between DENV and ZIKV is important, as simultaneous infection can increase the morbidity and mortality associated with these viral infections.

The initial infection with the DENV may be asymptomatic; however, depending on the stage of progression of the infection's progression, a considerable spectrum of clinical manifestations may arise. A classification for DENV has been established and delineated according to the advancement and intensity of the clinical manifestations of the infection. It is categorized into dengue with or without warning signs and severe dengue, which is characterized by alarming and acute symptoms, potentially resulting in functional impairment and multi-organ dysfunction [20]. DENV infections in the early stages can be asymptomatic in up to 80%, and about 20% present with an infection with symptoms in the early stage [21]. The clinical manifestations observed during the early phase of the pathology include mild fever, severe headache, retro-ocular pain, myalgia, arthralgia, rash, nausea, sore throat, and anorexia. Severe dengue infection, conversely, manifests symptoms that are frequently observed during the initial phase of the infection, including intense cephalalgia, retro-orbital discomfort, myalgia, arthralgia, cutaneous eruptions, nausea, and emesis. However, the symptoms that distinguish this progression are a sudden decrease in temperature, a decrease in blood pressure, persistent and severe abdominal pain, lethargy, hepatomegaly, and hemorrhages. It is important to mention that the progression of severe dengue infection can even lead to dengue shock syndrome with symptoms such as extreme pallor, high fever, a decrease in blood pressure, multiple organ damage, cardiorespiratory failure, respiratory distress, cardiac arrest, and altered consciousness [21, 22].

In relation to ZIKV, it is estimated that approximately 80% of infections may manifest as asymptomatic; however, when clinical manifestations occur, they tend to be predominantly mild in the majority of instances [23]. According to the clinical guide published by PAHO in 2022, pruritus is the most common symptom present in

ZIKV, while other works mention approximately 20 to 25% of cases with symptoms such as fever, rash, skin eruptions, arthralgia, myalgia, fatigue, headache, vomiting, diarrhea, eye redness, edema, abdominal pain, conjunctivitis, thrombocytopenia and hematospermia [19, 23]. In contrast, the presence of neurological symptoms derived from DENV and/or ZIKV infection has been reported. Guillain-Barré syndrome is the most common neurological symptom in ZIKV infections and severe stages of dengue. Other neurological symptoms associated with ZIKV include epilepsy, cerebral palsy, neuropathies, microcephaly, meningitis, and meningoencephalitis [24]. The association of Guillian-Barré syndrome and ZIKV was first observed in 2013–2014, due to cases presented in French Polynesia, and was subsequently reported in cases in the American continent in 2015–2016 [19]. The reported incidence of neurological damage due to DENV is approximately 5%; however, there is no consistent global incidence of neurological damage associated with DENV and ZIKV. Neuroinvasion by DENV was investigated using murine models, which resulted in the overexpression of proinflammatory cytokines IL-6, IL-10, TNF-α, IFN-γ, and MMP-9 in microglial cells [25]. In a study in Brazil, it was found that in women infected with ZIKV during pregnancy, approximately 80% of newborns were identified with manifestations of ZIKV, Congenital Syndrome, as well as microcephaly and other neurological alterations [26]. Due to the common symptoms, its diagnosis can be confused with other viral infections, such as dengue infection [16, 23].

In this regard, the differential diagnosis between dengue and Zika infections can be established based on clinical evaluation, considering the intensity and frequency of various signs and symptoms. In general, fever tends to be of moderate intensity in dengue infection, whereas in Zika infection, it is usually mild. A rash is an uncommon cutaneous manifestation in dengue, but it occurs frequently in Zika infection. Non-purulent conjunctivitis is a common feature of Zika infection but is infrequently observed in dengue cases. Other symptoms such as headache, myalgia, arthralgia, retro-orbital pain, and vomiting are typically more frequent and pronounced in dengue infection, while in Zika infection they tend to be milder or uncommon. These clinical features help guide presumptive diagnosis; however, confirmation through specific laboratory tests (RT-PCR-Reverse Transcriptase Polymerase Chain Reaction-, Detection of the NS1 viral protein, viral isolation, rapid immunochromatographic tests) is recommended due to the overlapping symptomatology of these viruses. On the other hand, according to PAHO/WHO, pharmacological interventions with paracetamol (acetaminophen) and metamizole are suggested for the signs and symptoms of mild forms of dengue and Zika; however, research into a specific treatment is still ongoing [1–3].

### 3.2 Clinical manifestations in patients with DENV and ZIKV coinfection

Coinfection occurs when a person is infected with both viruses, which are transmitted by the same mosquito vector. The clinical presentation of this coinfection can be complex, as both DENV and ZIKV infections share some symptoms, which can make diagnosis difficult. However, both diseases also have distinctive characteristics that facilitate their identification and diagnosis.

The clinical manifestations of DENV and ZIKV coinfection vary in severity. In mild to moderate cases, signs and symptoms include mild to moderate fever (around 39°C), moderate headache, myalgia, rash, sore throat, moderate cough, mild loss of appetite, fatigue, and lethargy. Severe cases may include a persistent, high fever (greater than 39.6°C), conjunctival hyperemia, retro-orbital pain, conjunctivitis, abdominal pain, frequent nausea and vomiting, dehydration, diarrhea, and a diffuse, pruritic maculopapular rash [27–29]. Among the most frequent symptoms in patients with DENV and ZIKV coinfection, fever of 39°C or higher, headache, myalgia, arthralgia, and rash have been reported [27].

Although reports on complications arising from coinfection by ZIKV and DENV are limited, cases have been documented that present manifestations such as thrombocytopenia, hepatomegaly, gingival and mucosal bleeding [30]. On the other hand, although there are few reports of pregnancy complications associated with this coinfection, cases of functional plagiocephaly in newborns and fetal death have been identified as possible consequences in women who presented coinfection with DENV and ZINV during pregnancy [30].

Currently, there is no specific treatment for DENV and ZIKV coinfection. Therefore, clinical management is based on symptomatic and supportive treatment similar to that used for individual infections, with the goal of alleviating symptoms. Coinfection with DENV and ZIKV represents a significant challenge for diagnosis and treatment due to the overlapping symptoms and complications associated with both viruses.

### 3.3 DENV and ZIKV coinfection in *in vitro* and mammalian models

Due to the cocirculation of both viruses in endemic regions, arbovirus coinfection has been the subject of multiple experimental studies, with the aim of understanding its dynamics of infection, dissemination, and viral transmission using cell and mammalian cultures.

Some authors report that a previous coinfection with one arbovirus can restrict or even prevent subsequent infections with another arbovirus [31]. However, recent research has yielded controversial and novel results depending on the type of infecting arbovirus, as well as the coinfection pattern. In this regard, several *in vitro* studies indicate that in C6/36HT cells, coinfection with DENV2 and DENV3 enhances DENV2 replication efficiency [32]. Furthermore, in the Aag2 cell line, it was shown that the Nicaraguan isolate of DENV2 NI-2B presented greater replication than the NI-1 isolate [33]. On the other hand, a study by Lin *et al.* indicates that viral replication in C6/36 cells (*Ae. albopictus*) sequentially infected with DENV strain 16,681 and ZIKV (Thai isolate), presents an increase in DENV2 viral replication compared to monoinfection; in contrast, ZIKV replication was drastically suppressed. Importantly, identifying the interaction between the ZIKV NS5 protein and the ssRNA of DENV2 suggests that there may be a cooperation in the use of cellular and molecular resources that contributes to and favors DENV2 viral replication [34]. Likewise, a subsequent study by Peng *et al.* presented that sequential coinfection at early times (12–36 h) with DENV2 (New Guinea strain C) and ZIKV (strain Z16006) in C6/36 cells did not influence the replication of either *Orthoflavivirus*; however, at late times of infection (48–72 h), ZIKV replication was observed to be suppressed. This suggests that infection with DENV2 could negatively interfere with ZIKV replication over time. In contrast, it was found that prior infection with ZIKV did not alter the replicative capacity of DENV2 in this cellular model [31].

Regarding animal models, Valiant *et al.* reported in rhesus macaques (*Macaca mulatta*) that sequential coinfection with Asian ZIKV genotype significantly enhances DENV2 viremia. However, in models with simultaneous coinfection, it has been observed that the viral loads in the macaques' plasma are very similar to monoinfection, suggesting that the replication kinetics of either virus are not modulated. However, this simultaneous coinfection is accompanied by a considerable activation of proinflammatory subpopulations of monocytes/macrophages and the release of numerous proinflammatory mediators, which have been associated with the appearance of the disease in its severe form [35].

## 4. DENV and ZIKV coinfection in the mosquito *Aedes aegypti*

Due to the current coexistence of DENV and ZIKV in endemic areas, their main mosquito vector, *Ae. aegypti*, has a greater chance of contracting simultaneous or mixed infections between these viruses. In this regard, several studies have focused on elucidating and understanding the mechanisms by which these complex coinfection interactions occur in the vector, and the effects they can cause. Various factors can impact mosquito susceptibility to coinfection, including the virus dose, which is directly proportional to the infection rate [36], the type of infecting virus [37], the mosquito's immune response [38], the pattern of coinfection (simultaneous and sequential) [39], as well as the genetic diversity and susceptibility of mosquitoes to virus infection and adaptation, where there is evidence indicating that Brazilian populations of *Ae. aegypti* mosquitoes are particularly more susceptible to DENV2 infections than African mosquito populations [40]. Dabo *et al.* found wide heterogeneity in the ability to acquire and replicate among different DENV strains and serotypes, concluding that this is determined by the specific association between the mosquito population and the DENV strain [41]. Likewise, Roundy *et al.* have reported that the Asian lineage of ZIKV is less infectious than the African lineage of ZIKV in *Ae. aegypti* populations present in the Americas [42]. Considering all these factors, three types of scenarios have been reported: the primary infection may enhance, inhibit, or have no effect on coinfections.

Furthermore, the coinfection phenomenon becomes more complex if we consider the existence of the four DENV serotypes; in this sense, *in silico* assays have modeled interactions between these findings differences considering the coinfection pattern [43]. Additionally, the analysis of simultaneous coinfection by different DENV serotypes in mosquitoes reveals that one serotype takes advantage and enhances its infection. For example, Quiner *et al.* studied the infectivity of different Nicaraguan DENV2 isolates and found that the NI-2B isolate had a replicative advantage over NI-1 up to 12 days after infection in *Ae. aegypti* [32]. Moreover, Quintero-Gil *et al.* found that in coinfected *Ae. aegypti* mosquitoes, the DENV2 serotype had a replication efficiency a thousand times greater than the DENV3 serotype at 4–14 days post-feeding [33]. Moreover, Vazeille *et al.* demonstrated that in *Ae. aegypti* mosquitoes from French Guiana, DENV4, have a competitive advantage over DENV1, enhancing its transmission 21 days after oral exposure to coinfected blood meal [44]. On the other hand, when infection occurs sequentially in the *Ae. aegypti* mosquito, it has been reported that the first exposure to one DENV serotype reduces the susceptibility and dissemination of the other. In this sense, Muturi *et al.* sequentially infected mosquitoes with DENV2 and DENV4 7 days after the first blood meal, demonstrating a decrease in susceptibility to secondary infection regardless of the serotype [45]. In conjunction with this, Serrato-Salas *et al.*, found that 7 days after challenging *Ae. aegypti* mosquitoes with an inactive version of DENV2 New Guinea C virus, were less susceptible to secondary DENV2 infection [46]. This evidence demonstrates that even DENV serotypes interact within the mosquito vector, and more than the type of infecting serotype, it seems that the coinfection pattern determines the behavior and survival of coexisting serotypes.

Within this complex context, we can also find coinfections between different *Orthoflavivirus*, mainly DENV and ZIKV, a product of the Zika pandemic in 2015, and from which constant subsequent outbreaks have occurred [36]. Likewise, several studies have evaluated coinfection in mosquitoes. In this regard, Lin *et al.* reported that by challenging *Ae. aegypti* mosquitoes (UGAL [University of Georgia Laboratory]/Rockefeller strain) with DENV2 (strain 16,681) and ZIKV (Thai isolate/1610acTw) sequentially, virus production and vector susceptibility to infection increased significantly [34]. Also, a study conducted in Manaus, the capital of the state of Amazonas, Brazil, showed that *Ae. aegypti* from this region is highly permissive to monoinfection and coinfection with DENV and ZIKV and is capable of co-transmitting both pathogens by bite. This coinfection significantly influences vector competence, favoring the transmission of ZIKV to the vertebrate host [44]. Furthermore, a study showed sequential coinfection of *Ae. aegypti* with Chikungunya virus (CHIKV), an alphavirus that is transmitted by the same vector, and ZIKV increased the transmission potential of ZIKV [47]. Similarly, Le Coupanec *et al.* reported that mixed CHIKV (La Reunion 06.21) and DENV (Bangkok, Thailand, 2BN32) infection facilitates viral replication in *Ae. aegypti* (Liverpool strain) [48]. All these findings support the hypothesis that a single mosquito bite has the potential to transmit two or more arboviruses simultaneously. They therefore reinforce the premise that simultaneous and/or sequential coinfection with arboviruses in mosquitoes increases their cotransmission, thus contributing to the spread and expansion of epidemics in humans. However, contrary to this evidence, Peng *et al.* presented that *Ae. aegypti* mosquitoes (Guangdong line) can be sequentially coinfected with DENV (New Guinea strain C) and ZIKV (strain Z16006), and regardless of the order of exposure to the virus, replication, dissemination and transmission potential of the secondary virus were inhibited, a phenomenon that was attributed to the activation of innate immunity in previously infected mosquitoes, altered lipid metabolism and an enhanced RNA interference (RNAi) pathway [31]. Complementing these findings, it has also been proposed that the presence of an arbovirus in *Aedes* mosquitoes does not appear to significantly influence the development of a secondary infecting virus. As indicated by the finding, initial infection with DENV2 (India, 803,347) or CHIKV (India, 061573) did not affect subsequent replication of ZIKV (Uganda, MR-766) or infection rates of the *Ae. aegypti* mosquito [49]. Another study showed that the coinfection of *Ae. aegypti* (from Poza Rica, Mexico) with ZIKV (from Puerto Rico, KU501215), CHIKV (from British Virgin Islands, KJ451624), and DENV2 (from Mérida, Mexico, AY449677) minimally affected vector competence, and that the vectors were able to transmit each viral pair, as well as all three viruses simultaneously [50].

These findings suggest that for coinfection to become established within a mosquito vector, several complex interactions are required. These interactions range from the type and dose of the infecting virus, the pattern of coinfection, the interaction between viruses, and the vector's immune response, among others. Ultimately, these interactions will be crucial and determine the infection potential of vertebrate hosts, primarily humans, in coendemic areas. Further experimental research is definitely needed to help us refine and understand the transmission dynamics of these *Orthoflaviviruses* in mosquito vectors.

## 5. Vectorial capacity and vector competence of DENV and ZIKV transmitted by *Aedes aegypti*

The relationship between *Ae. aegypti*, DENV and ZIKV determines vector competence. This interaction is linked to biotic and abiotic factors, which together define vector capacity and promote the spread of viruses among the human population in each time and space. Biotic factors are linked to the bionomics of the vector, that is, its survival time and life cycle, hematophagous habits, number of gonotrophic cycles (oviposition and fertility), biting rate (locomotor activity and host search), as well as population density and structure [51]. While abiotic factors include temperature, humidity, rainfall, population type (rural or urban), and the virus's own characteristics. Thus, vector competence refers to the ability and capacity of the *Ae. aegypti* vector to acquire the infection from a host infected with DENV and/or ZIKV and at the same time transmit it to other susceptible hosts [52].

Once the mosquito has acquired the infection, the virus replicates in the enterocytes of the midgut (MI), subsequently passing into the circulation until reaching the salivary glands to transmit the virus at the time of the bite or even to its offspring in each oviposition (vertical or transovarian transmission) [53–55]. In the context of DENV, interactions with the vector can be influenced by the agent's own characteristics, such as its availability in reservoirs, affinity for the vector or vectors present, replication capacity in the vector, pathogenicity, or the serotype of the virus [56]. Consequently, the susceptibility of the *Ae. aegypti* mosquito has a genetic basis related to the acquisition of DENV infection and its dissemination. Furthermore, the survival rate, the biting rate, and the blood-borne success rate have been described to increase transmission efficiency in urban areas, where population density is high compared to rural areas. As demonstrated in quasi-experimental studies, the survival rate is higher during the dry season compared to the rainy season while maintaining the ability to transmit DENV [57].

As delineated, female *Ae. aegypti* exhibits hematophagous and highly anthropophilic habits in both urban, peri-urban, and rural conditions due to their ability to domiciliate and adapt to the environment [56]. In this regard, Kamau *et al.* found that *Ae. aegypti* feeds primarily on human blood (51%) and observed opportunistic feeding on cattle, rodents, reptiles, and birds, while in rural areas, the human source was related to the degree of urbanization of the landscape [58]. Patterns like those reported by Sene *et al.* [59], where 78% of 1710 female *Ae. aegypti* collected inside and outside urban dwellings were fed human blood. The differences included blood from dogs, cats, horses, cattle, sheep, and rats. Regarding the number of blood meals, a second non-infectious blood meal was observed to significantly improve the competence of *Ae. aegypti* against DENV and ZIKV, shortening the extrinsic incubation period (EIP) and improving virus dissemination from the IM. This was evaluated through feeding on *Ae. aegypti* with a mixture of DENV, ZIKV, and defibrinated sheep blood, and subsequently analyzed at different times to determine the infection and transmission status [60]. Regarding feeding sites, Ouédraogo *et al.* [61], in their study with 1908 adult females, found that inside the home, a significant proportion of females fed with human blood with single ingestions were higher than 91% compared to outside.

On the other hand, regarding the relationship between mosquito size and vector capacity, it has been reported that the average size of female mosquitoes capable of seeking a blood source is larger than the rest of the newly emerged population, increasing the possibility of surviving a second gonotrophic cycle. Meanwhile, those of smaller-sized mosquitoes explore in search of a blood source more frequently and feed on blood several times during a gonotrophic cycle, increasing the possibility of acquiring and transmitting the viruses. In contrast, the study by Johnson *et al.* showed that the variation in the size of the DENV-infected blood meal is proportional to the size of the vector, also observing that in mosquitoes with a complete feed of DENV2-infected blood, the dissemination of the virus is earlier, and the EIP is shorter [62]. This has also been observed with ZIKV since, during the ingestion of infected blood, subsequent blood ingestion stimulates viral replication, increasing viral titer and improving dissemination to new susceptible hosts [63]. Similarly, the immune response and the vector microbiota are also related to the virus's ability to infect the female mosquito, as well as to the transmission mechanisms (horizontal or vertical-transovarial). Likewise, it has been described that the genetic characteristics of the vector populations are related to the intestinal microbiome and, in turn, to the infection rates by ZIKV and DENV, which impacts the ability to acquire, replicate, and transmit these viruses [64, 65].

Likewise, abiotic factors related to geographical characteristics such as climate, relief, hydrography, housing, pollution, type of population (rural, semi-urban, or urban), and human activity impact the virus-vector relationship [56]. In different regions of Latin America, due to the lack of planning for the growth and infrastructure of urban areas and growing human settlements, as well as the lack of services, such as the availability of water for human use, the increase in vector populations and consequently the spread of viruses is favored [66]. This was demonstrated in a research investigation conducted in Brazil, wherein geotechnological methods were employed to delineate various locations characterized by elevated population density, insufficient sanitation services, and diminished green spaces, detailing the epidemics that transpired during the summer and autumn seasons [67]. Similarly, variations have been observed between populations of *Ae. aegypti* from different regions and in their capacity to acquire and transmit ZIKV, possibly related to microclimatic factors [68]. Furthermore, microclimatic thermal variations directly influence the bionomics of the vector. High temperatures accelerate the development of life cycle stages, negatively impacting fertility rates, decreasing egg-laying success, and offspring viability [69, 70]. Research investigations aimed at assessing the impact of elevated temperatures on vectorial capacity regarding the transmission of ZIKV suggest a reduction in longevity [71]. Others have shown that vectorial competence for this virus is unimodal, observing that temperatures between 28 and 32°C are associated with high vectorial competence, while temperatures lower or higher than these are associated with low competence, a phenomenon that has also been observed in the transmission of DENV [72–74]. For its part, cooler temperatures improve the development of the vector's life stages and, consequently, vector competence, as has been evidenced in laboratory conditions, finding an effect on the shortening of the embryonic development period at 32°C, and on the larval and pupal stages at 37°C [75]. In this sense, microclimatic variations may be important in predicting transmission patterns based on seasonal thermal fluctuations, in which an increase in cases of these arboviruses has been observed during the summer [76].

## 6. Vertical transmission of DENV and ZIKV in *Aedes aegypti*

Both DENV and ZIKV have been shown to be able to be transmitted horizontally and vertically [77]. Horizontal transmission refers to the transmission of the virus to humans, and occurs when mosquitoes become infected by feeding on viremic hosts [78, 79]. This mode of transmission constitutes a significant mechanism contributing to the longevity and sustained presence of these members of the *Orthoflavivirus*.

On the other hand, vertical transmission is characterized by the transmission of these viruses to a proportion of their offspring (**Figure 2**), which also plays a crucial role in maintaining viruses within mosquito populations [78, 81]. In this regard, studies have shown that, in tropical regions, vertical transmission is more frequent in the vector *Ae. aegypti* compared to *Ae. albopictus* [82, 83]. It is known that, during infectious outbreaks, horizontal transmission between vectors predominates; on the other hand, during periods between epidemics, vertical transmission is a mechanism that these viruses use to their advantage to maintain their prevalence in nature. In line with the above, vertical transmission of DENV has been demonstrated during the inter-epidemic period, as observed in Mexico, where *Ae. aegypti* eggs were collected in the field, and the larvae hatched in the laboratory were analyzed by RT-PCR assays; this finding allowed the identification of serotypes DENV1, DENV2, and DENV3 [82]. On the other hand, under laboratory conditions, it was shown that DENV3 can be transmitted to up to seven consecutive generations of *Ae. aegypti*. However, an important observation was that, by the seventh generation, the mortality rate of infected larvae decreased significantly compared to the first two generations, demonstrating the prevalence of the virus in the vector and making it an ideal reservoir for its conservation [83]. In contrast, another study under laboratory conditions indicated that variations in the feeding of the *Ae. aegypti* mosquitoes do not affect the vertical transmission of DENV [84]. However, fieldwork carried out in Brazil, in areas with a predominance of DENV2 and DENV3 serotypes, revealed the presence of DENV2 in collected wild larvae and pupae [85]. Scientific evidence also confirms

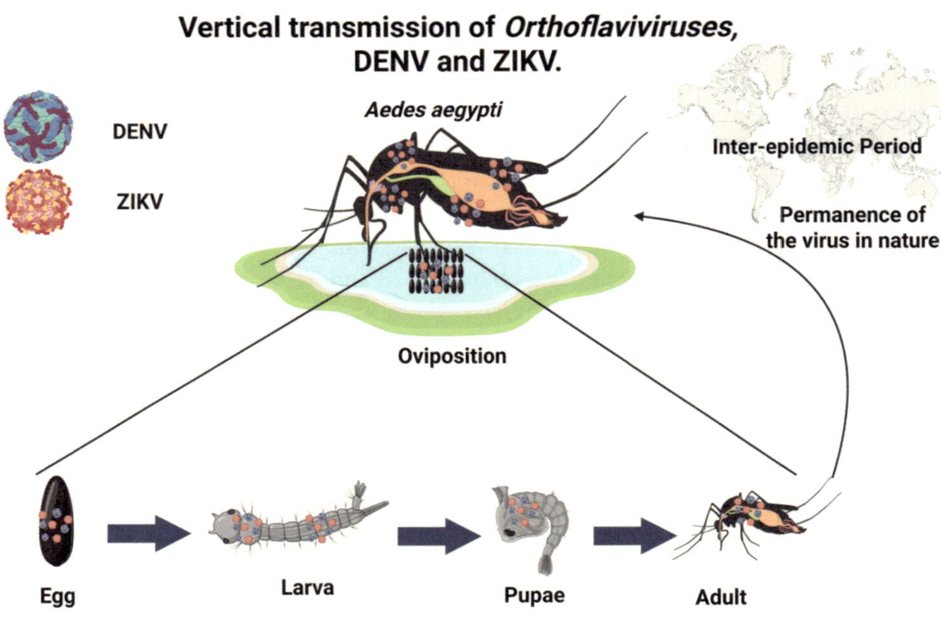


**Figure 2.**
*Vertical transmission of the* Orthoflaviviruses *DENV and ZIKV are described in the* Ae. aegypti *mosquito. During the inter-epidemic period, the virus finds a preservation mechanism through vertical transmission (VT), which can take place when the female mosquito is infected through hematophagy or during mating, where both viruses, having tropism for the ovaries, can be transmitted to the offspring and preserved in the different stages of the mosquito's development until its adult stage, promoting TV (based on [78, 80]. Created in BioRender.com.*

vertical transmission of DENV1 in *Ae. aegypti*, with an 11% rate recorded during egg laying [86]. Regarding DENV4, an RT-PCR study demonstrated its capacity for vertical transmission, both in single infections and in coinfections with DENV3, detecting viral genetic material in immature stages collected in the Philippines [87]. These data suggest that all four DENV serotypes have the potential for vertical transmission.

Regarding ZIKV, like DENV, it is primarily transmitted by *Ae. aegypti* and has also been shown to be able to be transmitted vertically to its offspring. This ability of the virus has been reported in Mexico and Brazil through RT-PCR and immunofluorescence assays, which have demonstrated the presence of ZIKV in both eggs collected in the field and their larvae hatched in the laboratory [88, 89]. However, in the case of the African lineage of ZIKV, vertical transmission has not been demonstrated; however, it is not ruled out that this lineage has the capacity to be transmitted to the offspring of the vector [78]. On the other hand, it has been reported that during vertical transmission of ZIKV, oviposition is affected by a decrease in the number of eggs but with an increase in their size, which suggests that this reduction could allow ZIKV-infected eggs to acquire more nutrients from the parental females [79]. Furthermore, vertical transmission of ZIKV can occur up to the fourth filial generation in the vector, a fact that demonstrates the reservoir capacity of *Ae. aegypti* for this virus [90]. In this sense, it is hypothesized that the vertical transmission mechanism is used by both DENV and ZIKV to overcome unfavorable environmental conditions. In this context, it has been reported that the replication of both viruses in the vector is affected by the environmental temperature in which the arthropod is located. However, studies have shown that vertical transmission of ZIKV and DENV can take place at various temperatures, 27, 30, and 33°C for ZIKV and 24–30°C for DENV [85, 91], supporting the idea that this is a preservation mechanism.

### 6.1 Vertical transmission mechanisms

Among the mechanisms for vertical transmission, it has been reported that male mosquitoes infected with DENV or ZIKV can transmit these viruses to females during mating. If the female is fertilized, both viruses show tropism toward the ovaries, which favors egg infection. This process can be enhanced if the female subsequently feeds on infected blood, thus increasing the likelihood that the eggs contain viral particles. As a result, vertical transmission of the virus to offspring occurs (**Figure 2**), culminating in vertical transmission in the progeny with an infection rate of 1:290–1:151 in the case of ZIKV and 1:6–1:4 in DENV [83, 85, 88, 92]. On the other hand, it has been shown that vertical transmission in a natural manner can occur under conditions of coinfection between DENV and ZIKV, although this appears less frequently (13%) compared to mono-infections in the vector (30%) [93].

Based on the aforementioned information, it is crucial to note the developmental cycle of *Ae. aegypti*, from egg to adult, requires approximately 12 days. During this time, the female mosquitoes, ready for oviposition, require blood feeding, a time when it can become vulnerable to acquiring DENV and/or ZIKV. In this context, the egg is one of the primary stages for serving as a reservoir for these *Orthoflaviviruses*, as it can withstand adverse weather conditions and prolonged periods of desiccation until favorable environmental factors, characterized by increased humidity and optimal thermal conditions, generally associated with rainfall or flooding, become favorable. Furthermore, in urbanized environments, artificial water reservoirs create favorable environments with the necessary conditions for the hatching of these eggs. Once the eggs hatch, the larval stage begins, which includes four instars that can last 6–8 days before entering the pupal stage [94, 95]. The pupal stage typically lasts approximately 1–2 days before the adult form emerges (**Figure 2**). Field studies have shown the presence of viruses in both the egg and pupal stages in both natural and laboratory environments, indicating that the persistence of both viruses can extend into the adult stage [88, 89]. Furthermore, female mosquitoes can lay around 300 eggs during multiple oviposition events, which maintains a higher frequency of viral circulation in natural ecosystems. Consequently, the egg and pupal stages, due to their potential reservoir capacity, deserve to be investigated and managed, as they could play a crucial role in eradicating the reservoirs of these two *Orthoflaviviruses* specifically. However, the use of synthetic chemicals is limited due to significant environmental impacts and adverse effects on human health [96]. For this reason, the proposal to use insecticides derived from natural sources emerges as an innovative approach for the management of this vector and, consequently, for reducing the transmission of both DENV and ZIKV.

## 7. Activities of plant extracts and essential oils against different life cycle forms of *Aedes aegypti*

Due to the absence of specific antiviral therapies for both pathologies and considering the historical prevalence of DENV epidemics alongside the current outbreaks of ZIKV, their management, and regulation have predominantly relied on the utilization of synthetic chemical insecticides. This reliance has, over time, resulted in the progressive development of resistance to various conventional chemical insecticides, in addition to revealing toxicological effects within the food chain and adversely impacting environmental quality [97], which in turn poses risks to both human and animal health. Therefore, it is necessary to increase knowledge of new strategies for mosquito vector control. In this context, there exists a considerable corpus of scholarly investigation focused on natural insecticides, particularly those sourced from botanical substances, owing to their diverse array of secondary metabolites that serve predominantly as a defense mechanism against their ecological adversaries. These compounds may confer benefits in terms of environmental safety, given that their residues are amenable to biodegradation and are non-toxic to non-target species [98–100]. Current research includes the analysis of botanical extracts and essential oils from various plants, evaluating *Ae. aegypti* from the early stages to the adult phase. Within the search for insecticides for *Ae. aegypti* control, the larvicidal test has been the most evaluated bioassay due to its ease and manipulation of the aquatic phase of the vector. For the adult phase, there are compounds that cause toxicity upon direct contact, as well as those with repellent action. This section aims to provide information on many of these plant extracts and essential oils, which have been tested as repellents and/or for their insecticidal activity against different life cycle stages of *Ae. aegypti*.

Based on the reviews by Silvério [101] and Priya *et al.* [102], we have an overview of previous works that, together with the analysis of recent works, help us to know more about the plants evaluated for their insecticidal effect during different stages of the life cycle of the *Ae. aegypti* mosquito, as well as their repellent action, in addition to identifying their geographical origin, allows us to identify that India has been the country with the most publications on botanical extracts, followed by Brazil, Indonesia, Madagascar, Saudi Arabia, and South Africa. Considering 19 plant families (Zingiberaceae, Aristolochiaceae, Asteraceae, Buddlejaceae, Fabaceae, Sapindaceae, Canellaceae, Cucurbitaceae, Apocynaceae, Rutaceae, Lamiaceae, Myristicaceae,

Amaryllidaceae, Lythraceae, Solanaceae, Rhammnaceae, Orchidaceae, Rubiaceae, and Sterculiacaeae), using from the whole plant or its aerial parts, bulbs, flowers, fresh or dried leaves, the root, root bark and seeds, being the dried leaves the most used part. Regarding the solvents used, water has been used, as well as a wide range of organic solvents, methanol being the most used solvent, followed by benzene, ethyl acetate, and n-hexane, and the least used were ethanol, chloroform, acetone, hexane, dichloromethane, hydroalcoholic, and ether. Crude extracts are obtained using various extraction methodologies employing organic solvents or water, while essential oils are obtained through steam distillation or hydrodistillation processes. The type of insecticidal activity (ovicidal, larvicidal, pupicidal, and adulticidal) is reported as mortality and lethal concentration (LC50 and LC90) values, along with egg hatching. Other evaluated activities include mosquito repellency and oviposition deterrence. **Table 1** provides some examples of the most recent studies evaluating these extracts and their efficacies at different stages of the mosquito vector's life cycle.

On the other hand, the use of essential oils has also been documented, and observed in 32 plant species belonging to 13 botanical families (Annonaceae, Apiaceae, Asteraceae, Cupressaceae, Euphorbiaceae, Lamiaceae, Myristicaceae, Myrtaceae, Poaceae, Rutaceae, Schisandraceae, Verbenaceae, and Zingiberaceae), with the Lamiaceae and Zingiberaceae families being the most represented in studies published between 2009 and 2025, this is because their multiple insecticidal effects on different stages of the life cycle of the vector mosquito have been reported. These species have been collected in 12 countries, mainly in Latin America (with special emphasis on Brazil) and Southeast Asia (particularly India, Thailand, and Vietnam). However, it was identified that there are few studies evaluating more than one biological property (such as larvicidal, ovicidal, pupicidal, or repellent activity), which represents a significant limitation when characterizing the comprehensive bioinsecticidal potential of these essential oils. **Table 2** presents examples of the main results of recent studies (2023–2025) that evaluated essential oils with more than one insecticidal activity against *Ae. aegypti*. Furthermore, it has been observed that most studies resort to the use of chromatographic techniques for the identification and isolation of active compounds [115, 116]. These compounds, once isolated, are usually evaluated individually to determine their biological activity. However, in many cases, the most promising results have been obtained using essential oils in their complete or unfractionated form, suggesting the existence of synergistic effects between their components. In this sense, the chemical composition of the essential oil, and therefore its effectiveness, is also influenced by the extraction technique used. Hydrodistillation is the most commonly reported extraction technique due to its simplicity and low cost. However, this methodology can induce thermal degradation of heat-sensitive compounds, thus altering the final chemical composition of the oil and, consequently, its biocidal efficacy. In contrast, some studies have begun to incorporate alternative extraction methods, such as supercritical fluid extraction, ultrasound-assisted extraction [117], and microwave-assisted extraction [118], which allow for higher yields and better preservation of bioactive compounds.

The present analysis allows us to understand that the insecticidal activity of the extracts/essential oils of leaves, roots, bulbs, seeds, fruits, and flowers exhibited significant activity against *Ae. aegypti*. Regarding the organic/aqueous extracts, the plant families with the greatest number of species tested against different stages of *Ae. aegypti* were Fabaceae, Asteraceae, Rubiaceae, and Ruteaceae. The ovicidal activity was assessed by exposing *Ae. aegypti* eggs to plant extracts for a period (approximately 24–28 hours), followed by counting the number of hatched/unhatched eggs

| *Plant species* | **Family** | **Country** | **Part used** | **Extraction solvent** | **Activity (*L1-L4)** | **Results** | **Time of analysis (h)** | **Reference** |
|---|---|---|---|---|---|---|---|---|
| *Helicteres velutina* | *Sterculia-caeae* | Brazil | Aerial parts | Hexane | Ovicide | Zero hatchability at 5 μg/mL | 25 days | [103] |
| | | | | Dichlorometane | Ovicide | Zero hatchability at 5 μg/mL | 25 days | |
| | | | | Hexane | Pupicide | LC50 0.12 mg/mL LC90 0.14 mg/mL | 24 | |
| | | | | Dichlorometane | Pupicide | LC50 8.85 mg/mL LC90 26.93 mg/mL | 24 | |
| | | | | Hexane | Adulticide | LC50 8.01 mg/mL LC90 36.14 mg/mL | 24 | |
| | | | | Dichlorometane | Adulticide | LC50 0.74 mg/mL LC90 43.89 mg/mL | 24 | |
| | | | | Hexane | Repellent | 100% of repellency at 3.88 mg/mL | 2 | |
| | | | | Dichlorometane | Repellent | 100% of repellency at 5.80 mg/mL | 2 | |
| *Ateleia glazioveana* | *Fabaceae* | Brazil | Leaves | Dichloromethane | Larvicide (L3) | LC50 95.4 μg/mL | 24 | [104] |
| | | | | Dichloromethane | Repellent | 85.71% of repellency by cream based on ichloromethane 5% | ND | |
| | | | | hydroalcoholic | Repellent | 57.14% of repellency by cream based on ichloromethane 5% | ND | |
| | | | | Ethanol | Larvicide (L3) | LC50 0.72 mg/mL LC90 16.7 mg/mL | 24 | [105] |
| *Blumea lacera* | *Asteraceae* | India | Leaves | Ethanol | Ovicide | 10% hatchability at 0.3 mg | 48 | |
| | | | | Ethanol | Larvicide (L3) | LC50 7.349 mg/mL LC90 1264.37 mg/mL | 24 | |
| *Neanotis montholonii* | Rubiaceae | India | Leaves | Ethanol | Ovicide | 25% hatchability at 0.3 mg | 48 | |
| | | | | Ethanol | Larvicide (L3) | — | 24 | |
| *Neanotis lancifolia* | Rubiaceae | India | Leaves | Ethanol | Ovicide | 30% hatchability at 0.3 mg | 48 | |
| | | | | Ethanol | Larvicide (L3) | — | 24 | |
| *Peltophorum pterocarpum* | *Fabaceae* | India | Leaves | n-Hexane | Ovicide | 24% hatchability at 500 ppm | 120 | [106] |

| *Plant species* | **Family** | **Country** | **Part used** | **Extraction solvent** | **Activity (*L1-L4)** | **Results** | **Time of analysis (h)** | **Reference** |
|---|---|---|---|---|---|---|---|---|
| | | | | Methanol | Ovicide | 6.4% hatchability at 500 ppm | 120 | |
| | | | | Chloroform | Ovicide | 24% hatchability at 500 ppm | 120 | |
| | | | | n-Hexane | Larvicide (L3) | LC50 219.14 ppm LC90 1120.27 ppm | 24 | |
| | | | | Methanol | Larvicide (L3) | LC50 111.77 ppm LC90 330.71 ppm | 24 | |
| | | | | Chloroform | Larvicide (L3) | LC50 170.74 ppm LC90 684.06 ppm | 24 | |
| | | | | n-Hexane | Pupicide | LC50 303.07 ppm LC90 1195.92 ppm | 24 | |
| | | | | Methanol | Pupide | LC50 235.40 ppm LC90 1011.83 ppm | 24 | |
| | | | | Chloroform | Pupicide | LC50 226.01 ppm LC90 1155.82 ppm | 24 | |
| *Peltophorum pterocarpum* | *Fabaceae* | India | Flowers | n-Hexane | Ovicide | 9.6% hatchability at 500 ppm | 120 | [107] |
| | | | | Methanol | Ovicide | 3.2% hatchability at 500 ppm | 120 | |
| | | | | Chloroform | Ovicide | 5% hatchability at 500 ppm | 120 | |
| | | | | n-Hexane | Larvicide (L3) | LC50 111.81 ppm LC90 330.74 ppm | 24 | |
| | | | | Methanol | Larvicide (L3) | LC50 121.92 ppm LC90 371.14 ppm | 24 | |
| | | | | Chloroform | Larvicide (L3) | LC50 357.2 ppm LC90 982.1 ppm | 24 | |
| | | | | n-Hexane | Pupicide | LC50 3247.12 ppm LC90 18226.7 ppm | 24 | |
| | | | | Methanol | Pupicide | LC50 172.8 ppm LC90 791.30 ppm | 24 | |
| | | | | Chloroform | Pupicide | LC50 425.8 ppm LC90 1218.2 ppm | 24 | |
| *Coffea canephora* | Rubiaceae | India | Leaves | Hexane | Ovicide | 30.4% hatchability at 500 ppm | 120 | [107] |
| | | | | Chloroform | Ovicide | 3.2% hatchability at 500 ppm | 120 | |
| | | | | Methanol | Ovicide | 3.2% hatchability at 500 ppm | 120 | |
| | | | | Hexane | Larvicide (L3) | LC50 331.5 ppm LC90 1104.3 ppm | 24 | |

| *Plant species* | Family | Country | Part used | Extraction solvent | Activity (*L1-L4) | Results | Time of analysis (h) | Reference |
|---|---|---|---|---|---|---|---|---|
| | | | | Chloroform | Larvicide (L3) | LC50 353.0 ppm LC90 1138.0 ppm | 24 | |
| | | | | Methanol | Larvicide (L3) | LC50 116.2 ppm LC90 258.5 ppm | 24 | |
| | | | | Hexane | Pupicide | LC50 3793.8 ppm LC90 33300.8 ppm | 24 | |
| | | | | Chloroform | Pupicide | LC50 4089.2 ppm LC90 26161.1 ppm | 24 | |
| | | | | Methanol | Pupicide | LC50 261.7 ppm LC90 746.7 ppm | 24 | |

*The larval phase of* Ae. aegypti *is the period of feeding and growth and consists of a cycle of four larval stages (L1, L2, L3, and L4).*

**Table 1.**
*The larvicidal, pupicidal, ovicidal, adulticidal, repellent, and oviposition activities of organic/aqueous extracts against the* Ae. aegypti *mosquito.*

| Plant species | Family | Country | Part used | Activity (*L1-L4) | Results | Time of analysis (h) | Reference |
|---|---|---|---|---|---|---|---|
| *Bocageopsis multiflora (Mart.)* | Annonaceae | Brasil | Leaves | Larvicidal | LC50 40.8 ppm; CL90 71.9 ppm | 24 | [108] |
| | | | | Adulticidal | LC50 12.5 ppm; CL90 17.1 ppm | 1.5 | |
| *Brassica nigra* | | India | Seed | Larvicidal | LC50 23.2 μL/L | 24 | [109] |
| | | | | Pupicidal | LC50 26.5 μL/L | 24 | |
| | | | | Ovicidal | 100% ovicidal effect | 24 | |
| *Citrus reticulata* | Rutaceae | Myanmar | Peels | Larvicidal L3/L4 | LC50 0.0015 g LC90 0.0051 g | 24 | [110] |
| | | | | Repellency | 100% repellency al 120 min | 4 | |
| | | | | Ovicidal | 100% ovicidal effect for 4 days at 0.01 g EO | 96 | |
| *Helianthus annuus (L.)* | *Asteraceae* | Nigeria | Seed | Larvicidal (L4) | LC50 2021.32 ppm | 0.75 | [111] |
| | | | | Adulticidal | LC50 2565.20 ppm | 0.75 | |
| *Ocimum basilicum* | Lamiaceae | Brasil | Leaves | Repellency | 70.5% repellency | 24 | |
| | | Malasia | Basil oil | Repellency | 98.9% to 2.22% repellency for 5 h | 6 | |
| | | Brasil | Leaves | Larvicidal (L4) | LC50 42.15 ppm; 50.35 ppm | 24 | [104, 112] |
| | | | Leaves | Larvicidal (L4) | LC50 40.94 ppm; 54.26 ppm | 48 | |
| *Origanum minutiflorum* | Lamiaceae | Turquia | Aerial part | Larvicidal | LC50 24.5 ppm | 120 | [113] |
| | | | | Ovicidal | no ovicidal effects were detected | 240 | |
| *Salvia dorystoechas* | Lamiaceae | Turquia | Aerial part | Larvicidal | LC50 100.6 ppm | 120 | [114] |
| | | | | Ovicidal | no ovicidal effects were detected | 240 | |

**The larval phase of* Ae. aegypti *is the period of feeding and growth, and consists of a cycle of four larval stages (L1, L2, L3, and L4).*

**Table 2.**
*The larvicidal, pupicidal, ovicidal, adulticidal, repellent, and oviposition activities of essential oil against the* Ae. aegypti *mosquito.*

(mortality) compared to the number of eggs laid. Larvicidal activity was measured by lethal concentration (LC) and larval mortality (represented in mg/mL, µg/mL, or ppm). The assessment of the pupicidal activity of the plant extracts/essential oils was inferred based on the mortality rates of *Ae. aegypti* pupae when exposed to these for a period of approximately 24 hours, and the lethal concentration and lethal time values were calculated. The adulticidal activity of the plant extracts and essential oils was assessed by determining the percentage mortality of adult mosquitoes, lethal concentrations, and lethal times when exposed to these extracts and essential oils for a period of 24 hours. For oviposition deterrence, the extracts were exposed to adult female *Ae. aegypti* mosquitoes, and the percentage of repellency was measured based on the number of eggs laid by the mosquitoes. Repellency tests were performed with animal models or human volunteers, exposing them to adult female *Ae. aegypti* mosquitoes at different time intervals and concentrations, and the percentage of landing inhibition was ultimately calculated.

Furthermore, these analyzed works allow us to notice that from 2009 to date, few studies have evaluated the insecticidal effect in more than two phases of the life cycle of the mosquito vector, the most analyzed phases being the ovicidal-larvicidal activity or the ovicidal-pupicidal activity, in addition to denoting the lower integration of the repellent activity, which highlights the importance of characterizing the functionality of each plant in its insecticidal activity for each phase of the mosquito in addition to its repellent function, this with the objective of increasing the information in this regard, as well as enhancing the options for the use of natural insecticides, emphasizing and giving importance to the fact that in the early stages of the mosquito's life they play an essential role, not only for the control of the population density of the vectors, but also in reducing the risk of viral transmission, due to the transovarial transmission characteristic of DENV and ZIKV, which increases the possibility of contagion in human populations. In this sense, the great diversity of plants with toxic effects on the different life stages of the *Ae. aegypti* mosquitoes demonstrate their potential to be used as a feasible alternative to synthetic chemical insecticides, given their lower environmental impact, which, together with their lower toxicity to non-target insect species, make them a considerable option. However, more extensive research is needed to determine the type of plant, the type of extract, the type of metabolite, the concentration, and the optimal combinations to enhance the use of these natural insecticides for vector control.

## 8. Conclusions

While *Orthoflaviviruses* such as DENV and ZIKV represent a distinct category of arboviruses, their unique attributes and transmission mechanisms underscore the intricate nature of viral interactions within ecological systems. In this context, the transmission cycles of DENV and ZIKV (both in sylvatic and urban areas) contribute significantly to the spread of the disease and the high prevalence rates each year. The simultaneous circulation of DENV and ZIKV in tropical regions represents a significant challenge for public health. Although both infections can be asymptomatic or cause mild symptoms, they can also lead to severe illness with neurological consequences in offspring or pregnancy complications. Coinfection with DENV and ZIKV has been little studied; however, it can increase morbidity and clinical complexity, requiring more specialized medical care. *In vitro* coinfection studies have revealed to us the effects behind replication: on the one hand, enhancing DENV replication in the presence of ZIKV, and on the other, suggesting cooperation between viral elements of both viruses to promote replication. Moreover, the establishment of coinfection with DENV and ZIKV within a mosquito vector necessitates the existence of various intricate interactions (dose and strains or serotypes of infecting virus, the coinfection pattern, the interaction between viruses, the immune response of the vector, among others) that will be crucial and will determine the infection potential in vertebrate hosts, mainly humans, in coendemic areas. Therefore, understanding the mechanisms by which the acquisition, replication, and transmission of DENV and ZIKV occur by the *Ae. aegypti* mosquito in the susceptible population, that is, its vectorial capacity is crucial to understanding dispersion in time and space. Within these interactions, vertical transmission in the *Ae. aegypti* vector takes on vital importance as a mechanism of viral preservation, since, during oviposition, the viability of infected eggs could persist in an adverse environment, and once optimal conditions are found, development to the adult stage will be encouraged, thus maintaining the prevalence of both viruses. Above all, it is important to understand the behavior of the vector mosquito and its adaptive capacity to indoor environments, primarily due to the hematophagous habit of female mosquitoes. However, additional factors such as climate change, population movements, global trade, uncontrolled urbanization, and water shortages, among others, have allowed the vector to establish itself in new regions of the world, allowing the spread of these *Orthoflaviviruses*. This could be improved by controlling the various stages of insect development through the implementation and use of natural products of plant origin as an alternative to chemical insecticides. Ecological botanical insecticides are an innovative approach to control *Ae. aegypti* mosquito populations and therefore of DENV and ZIKV transmission, in addition to being safer for the environment and posing a lower risk to human health.

## Conflict of interest

The authors declare no conflict of interest.

## Author details

Josselin Carolina Corzo-Gómez[1*], Mayra Montecillo-Aguado[2], Esther Candelaria Martínez-Molina[1], Sury Antonio López-Cancino[3], Victor Javier Cruz-Holguin[4,5], Giovani Visoso-Carvajal[6], Alejandra García-Aguilar[6,7], María Fernanda García-Aguilar[8] and Sandra Paola Martínez-Frías[4]

1 School of Chemical Sciences, Autonomous University of Chiapas, Ocozocoautla de Espinosa, Chiapas, Mexico

2 Translational Biomedicine Laboratory, Division of Health Sciences, University of Guanajuato, Leon, Guanajuato, Mexico

3 Head of Health Research Office, IMSS BIENESTAR State Coordination Chiapas, Health Services of the Mexican Institute of Social Security for Welfare (IMSS BIENESTAR), Tuxtla Gutiérrez, Mexico

4 Research and Advanced Studies Center of the National Polytechnic Institute (Cinvestav), Mexico City, Mexico

5 National Institute of Perinatology (INPer), Mexico City, Mexico

6 Conservation Medicine Laboratory, School of Medicine of the National Polytechnic Institute, Mexico City, Mexico

7 Neuropharmacology Laboratory, Higher School of Medicine of the National Polytechnic Institute, Mexico City, Mexico

8 Graduate Studies and Research Division (DEPei) in Dentistry at the National Autonomous University of Mexico (UNAM), Mexico City, Mexico

*Address all correspondence to: josselin.corzo@unach.mx; josselin.corzo@gmail.com

## References

[1] Rabaan AA, Bazzi AM, Al-Ahmed SH, Al-Ghaith MH, Al-Tawfiq JA. Overview of Zika infection, epidemiology, transmission and control measures. Journal of Infection and Public Health. 2017;**10**(2):141-149

[2] Roy SK, Bhattacharjee S. Dengue virus: Epidemiology, biology, and disease etiology. Canadian Journal of Microbiology. 2021;**67**(10):687-702

[3] Pourzangiabadi M, Najafi H, Fallah A, Goudarzi A, Pouladi I. Dengue virus: Etiology, epidemiology, pathobiology, and developments in diagnosis and control – A comprehensive review. Infection, Genetics and Evolution. 2025;**127**:105710

[4] Khanam A, Gutiérrez-Barbosa H, Lyke KE, Chua JV. Immune-mediated pathogenesis in dengue virus infection. Viruses. 2022;**14**(11):2575

[5] Halsey ES, Marks MA, Gotuzzo E, Fiestas V, Suarez L, Vargas J, et al. Correlation of serotype-specific dengue virus infection with clinical manifestations. PLoS Neglected Tropical Diseases. 2012;**6**(5):e1638

[6] Sinha S, Singh K, Ravi Kumar YS, Roy R, Phadnis S, Meena V, et al. Dengue virus pathogenesis and host molecular machineries. Journal of Biomedical Science. 2024;**31**(1):43

[7] Osawa T, Aoki M, Ehara H, Sekine S, ichi. Structures of dengue virus RNA replicase complexes. Molecular Cell. 2023;**83**(15):2781-2791.e4

[8] Hills SL, Fischer M, Petersen LR. Epidemiology of Zika virus infection. The Journal of Infectious Diseases. 2017;**216**(Suppl. 10):S868-S874

[9] Ferraris P, Yssel H, Missé D. Zika virus infection: An update. Microbes and Infection. 2019;**21**(8-9):353-360

[10] Kong W, Li H, Zhu J. Zika virus: The transboundary pathogen from mosquito and updates. Microbial Pathogenesis. 2018;**114**:476-482

[11] Dias ÍKR, CL da S S, Martins RMG, Santana KFS, Lopes M d SV, Joventino ES, et al. Zika virus: - a review of the main aspects of this type of arbovirosis. Revista da Sociedade Brasileira de Medicina Tropical. 2018;**51**(3):261-269

[12] Hu T, Li J, Carr MJ, Duchêne S, Shi W. The Asian lineage of Zika virus: Transmission and evolution in Asia and the Americas. Virologica Sinica. 2019;**34**(1):1-8

[13] Li HH, Su MP, Wu SC, Tsou HH, Chang MC, Cheng YC, et al. Mechanical transmission of dengue virus by Aedes aegypti may influence disease transmission dynamics during outbreaks. eBioMedicine. 2023;**94**:104723

[14] Vue D, Tang Q. Zika virus overview: Transmission, origin, pathogenesis, animal model and diagnosis. Zoonoses. 2021;**1**(1). DOI: 10.15212/zoonoses-2021-0017

[15] Chan M, Johansson MA. The incubation periods of dengue viruses. PLoS One. 2012;**7**(11):e50972

[16] Pielnaa P, Al-Saadawe M, Saro A, Dama MF, Zhou M, Huang Y, et al. Zika virus-spread, epidemiology, genome, transmission cycle, clinical manifestation, associated challenges, vaccine and antiviral drug development. Virology. 2020;**543**:34-42

[17] Vasilakis N, Cardosa J, Hanley KA, Holmes EC, Weaver SC. Fever from the forest: Prospects for the continued emergence of sylvatic dengue virus and its impact on public health. Nature Reviews. Microbiology. 2011;**9**(7):532-541

[18] Rudolph KE, Lessler J, Moloney RM, Kmush B, Cummings DAT. Incubation periods of mosquito-borne viral infections: A systematic review. The American Society of Tropical Medicine and Hygiene. 2014;**90**(5):882-891

[19] Song BH, Yun SI, Woolley M, Lee YM. Zika virus: History, epidemiology, transmission, and clinical presentation. Journal of Neuroimmunology. 2017;**308**:50-64

[20] Pang X, Zhang R, Cheng G. Progress towards understanding the pathogenesis of dengue hemorrhagic fever. Virologica Sinica. 2017;**32**(1):16-22

[21] Khan MB, Yang ZS, Lin CY, Hsu MC, Urbina AN, Assavalapsakul W, et al. Dengue overview: An updated systemic review. Journal of Infection and Public Health. 2023;**16**(10):1625-1642

[22] Parveen S, Riaz Z, Saeed S, Ishaque U, Sultana M, Faiz Z, et al. Dengue hemorrhagic fever: A growing global menace. Journal of Water and Health. 2023;**21**(11):1632-1650

[23] Plourde AR, Bloch EM. A literature review of Zika virus. Emerging Infectious Diseases. 2016;**22**(7):1185-1192

[24] Santos LLM, de Aquino EC, Fernandes SM, Ternes YMF, Feres VC d R. Dengue, chikungunya, and Zika virus infections in Latin America and the Caribbean: A systematic review. Revista Panamericana de Salud Pública. 2023;**47**:1

[25] Fong SL, Wong KT, Tan CT. Dengue virus infection and neurological manifestations: An update. Brain. 2024;**147**(3):830-838

[26] Ba F, Li H, Ding S, Guo T, Zhang J, Sun Y. Zika virus-related birth defects and neurological complications: A systematic review and meta-analysis. Reviews in Medical Virology. 2025;**35**(2):e70019

[27] Estofolete CF, Terzian ACB, Colombo TE, de Freitas GG, Ferraz HC, da Silva RA, et al. Co-infection between Zika and different dengue serotypes during DENV outbreak in Brazil. Journal of Infection and Public Health. 2019;**12**(2):178-181

[28] Chia PY, Sen YH, Ho H, Chow A, Sadarangani SP, Chan M, et al. Clinical features of patients with Zika and dengue virus co-infection in Singapore. The Journal of Infection. 2017;**74**(6):611-615

[29] Lovine NM, Lednicky J, Cherabuddi K, Crooke H, White SK, Loeb JC, et al. Coinfection with Zika and Dengue-2 viruses in a Traveler returning from Haiti, 2016: Clinical presentation and genetic analysis. Clinical Infectious Diseases. 2017;**64**(1):72-75

[30] Lobkowicz L, Ramond A, Sanchez Clemente N, Ximenes RA d A, Miranda-Filho D d B, Montarroyos UR, et al. The frequency and clinical presentation of Zika virus coinfections: A systematic review. BMJ Global Health. 2020;**5**(5):e002350

[31] Peng J, Zhang M, Wang G, Zhang D, Zheng X, Li Y. Biased virus transmission following sequential coinfection of Aedes aegypti with dengue and Zika viruses. PLoS Neglected Tropical Diseases. 2024;**18**(4):e0012053

[32] Quintero-Gil DC, Ospina M, Osorio-Benitez JE, Martinez-Gutierrez M.

Differential replication of dengue virus serotypes 2 and 3 in coinfections of C6/36 cells and Aedes aegypti mosquitoes. The Journal of Infection in Developing Countries. 2014;**8**(07):876-884

[33] Quiner CA, Parameswaran P, Ciota AT, Ehrbar DJ, Dodson BL, Schlesinger S, et al. Increased replicative fitness of a dengue virus 2 clade in native mosquitoes: Potential contribution to a clade replacement event in Nicaragua. Journal of Virology. 2014;**88**(22):13125-13134

[34] Lin DCD, Weng SC, Tsao PN, Chu JJH, Shiao SH. Co-infection of dengue and Zika viruses mutually enhances viral replication in the mosquito Aedes aegypti. Parasites & Vectors. 2023;**16**(1):160

[35] Valiant WG, Mattapallil MJ, Higgs S, Huang YJS, Vanlandingham DL, Lewis MG, et al. Simultaneous coinfection of macaques with Zika and dengue viruses does not enhance acute plasma viremia but leads to activation of monocyte subsets and biphasic release of pro-inflammatory cytokines. Scientific Reports. 2019;**9**(1):7877

[36] Souza-Neto JA, Powell JR, Bonizzoni M. Aedes aegypti vector competence studies: A review. Infection, Genetics and Evolution. 2019;**67**:191-209

[37] Kobayashi D, Kai I, Faizah AN, Moi ML, Tajima S, Takasaki T, et al. Comparative analysis of the susceptibility of Aedes aegypti and Japanese Aedes albopictus to all dengue virus serotypes. Tropical Medicine and Health. 2023;**51**(1):61

[38] Ramirez JL, Souza-Neto J, Torres Cosme R, Rovira J, Ortiz A, Pascale JM, et al. Reciprocal tripartite interactions between the Aedes aegypti midgut microbiota, innate immune system and dengue virus influences vector competence. PLoS Neglected Tropical Diseases. 2012;**6**(3):e1561

[39] Chaves BA, Orfano AS, Nogueira PM, Rodrigues NB, Campolina TB, Nacif-Pimenta R, et al. Coinfection with Zika virus (ZIKV) and dengue virus results in preferential ZIKV transmission by vector bite to vertebrate host. The Journal of Infectious Diseases. 2018;**218**(4):563-571

[40] Gonçalves CM, Melo FF, Bezerra JM, Chaves BA, Silva BM, Silva LD, et al. Distinct variation in vector competence among nine field populations of Aedes aegypti from a Brazilian dengue-endemic risk city. Parasites & Vectors. 2014;7(1):320

[41] Dabo S, Henrion-Lacritick A, Lecuyer A, Jiolle D, Paupy C, Ayala D, et al. Extensive variation and strain-specificity in dengue virus susceptibility among African Aedes aegypti populations. PLoS Neglected Tropical Diseases. 2024;**18**(3):e0011862

[42] Roundy CM, Azar SR, Rossi SL, Huang JH, Leal G, Yun R, et al. Variation in *Aedes aegypti* Mosquito competence for Zika virus transmission. Emerging Infectious Diseases. 2017;**23**(4):625-632

[43] Kang DS, Alcalay Y, Lovin DD, Cunningham JM, Eng MW, Chadee DD, et al. Larval stress alters dengue virus susceptibility in Aedes aegypti (L.) adult females. Acta Tropica. 2017;**174**:97-101

[44] Vazeille M, Gaborit P, Mousson L, Girod R, Failloux AB. Competitive advantage of a dengue 4 virus when co-infecting the mosquito Aedes aegypti with a dengue 1 virus. BMC Infectious Diseases. 2016;**16**(1):318

[45] Muturi EJ, Buckner E, Bara J. Superinfection interference between

dengue-2 and dengue-4 viruses in *Aedes aegypti* mosquitoes. Tropical Medicine & International Health. 2017;**22**(4):399-406

[46] Serrato-Salas J, Izquierdo-Sánchez J, Argüello M, Conde R, Alvarado-Delgado A, Lanz-Mendoza H. Aedes aegypti antiviral adaptive response against DENV-2. Developmental and Comparative Immunology. 2018;**84**:28-36

[47] Magalhaes T, Robison A, Young M, Black W, Foy B, Ebel G, et al. Sequential infection of Aedes aegypti mosquitoes with chikungunya virus and Zika virus enhances early Zika virus transmission. Insects. 2018;**9**(4):177

[48] Le Coupanec A, Tchankouo-Nguetcheu S, Roux P, Khun H, Huerre M, Morales-Vargas R, et al. Co-infection of mosquitoes with chikungunya and dengue viruses reveals modulation of the replication of both viruses in midguts and salivary glands of Aedes aegypti mosquitoes. International Journal of Molecular Sciences. 2017;**18**(8):1708

[49] Mourya DT, Gokhale MD, Majumdar TD, Yadav PD, Kumar V, Mavale MS. Experimental Zika virus infection in Aedes aegypti. The Indian Journal of Medical Research. 2018;**147**(1):88-96

[50] Rückert C, Weger-Lucarelli J, Garcia-Luna SM, Young MC, Byas AD, Murrieta RA, et al. Impact of simultaneous exposure to arboviruses on infection and transmission by Aedes aegypti mosquitoes. Nature Communications. 2017;**8**(1):15412

[51] Mateescu I, Lequime S. Dengue-mediated changes in the vectorial capacity of *Aedes aegypti* (Diptera: Culicidae): Manipulation of transmission or infection by-product? Journal of Medical Entomology. 2025;**62**(1):19-28

[52] Liu-Helmersson J, Stenlund H, Wilder-Smith A, Rocklöv J. Vectorial capacity of Aedes aegypti: Effects of temperature and implications for global dengue epidemic potential. PLoS One. 2014;**9**(3):e89783

[53] Lewis J, Gallichotte EN, Randall J, Glass A, Foy BD, Ebel GD, et al. Corrigendum: Intrinsic factors driving mosquito vector competence and viral evolution: A review. Frontiers in Cellular and Infection Microbiology. 2024;**13**:14

[54] Ciota AT, Chin PA, Ehrbar DJ, Micieli MV, Fonseca DM, Kramer LD. Differential effects of temperature and mosquito genetics determine transmissibility of arboviruses by Aedes aegypti in Argentina. The American Journal of Tropical Medicine and Hygiene. 2018;**99**(2):417-424

[55] Ahebwa A, Hii J, Neoh KB, Chareonviriyaphap T. Aedes aegypti and Aedes albopictus (Diptera: Culicidae) ecology, biology, behaviour, and implications on arbovirus transmission in Thailand: Review. One Health. 2023;**16**:100555

[56] Kramer LD, Ciota AT. Dissecting vectorial capacity for mosquito-borne viruses. Current Opinion in Virology. 2015;**15**:112-118

[57] Hugo LE, Jeffery JAL, Trewin BJ, Wockner LF, Thi Yen N, Le NH, et al. Adult survivorship of the dengue mosquito Aedes aegypti varies seasonally in Central Vietnam. PLoS Neglected Tropical Diseases. 2014;**8**(2):e2669

[58] Kamau WW, Sang R, Rotich G, Agha SB, Menza N, Torto B, et al. Patterns of Aedes aegypti abundance, survival, human-blood feeding and relationship with dengue risk, Kenya. Frontiers in Tropical Diseases.

2023;**4**:1113531. DOI: 10.3389/fitd.2023.1113531

[59] Sene NM, Diouf B, Gaye A, Ndiaye EH, Ngom EHM, Gueye A, et al. Blood feeding patterns of Aedes aegypti populations in Senegal. The American Journal of Tropical Medicine and Hygiene. 2022;**106**(5):1402-1405

[60] Armstrong PM, Ehrlich HY, Magalhaes T, Miller MR, Conway PJ, Bransfield A, et al. Successive blood meals enhance virus dissemination within mosquitoes and increase transmission potential. Nature Microbiology. 2019;**5**(2):239-247

[61] Ouédraogo WM, Zanré N, Sombié A, Yameogo F, Gnémé A, Sanon A, et al. Blood-feeding patterns and resting behavior of Aedes aegypti from three health districts of Ouagadougou City, Burkina Faso. The American Journal of Tropical Medicine and Hygiene. 2024;**111**(6):1295-1301

[62] Johnson RM, Cozens DW, Ferdous Z, Armstrong PM, Brackney DE. Increased blood meal size and feeding frequency compromise Aedes aegypti midgut integrity and enhance dengue virus dissemination. PLoS Neglected Tropical Diseases. 2023;**17**(11):e0011703

[63] Cui Y, Megawati D, Lin J, Rehard DG, Grant DG, Liu P, et al. Cytoskeleton-associated gelsolin responds to the midgut distention process in saline meal-fed *Aedes aegypti* and affects arbovirus dissemination from the midgut. The FASEB Journal. 2024;**38**(14):e23764. DOI: 10.1096/fj.202302684RR

[64] Hegde S, Rasgon JL, Hughes GL. The microbiome modulates arbovirus transmission in mosquitoes. Current Opinion in Virology. 2015;**15**:97-102

[65] Accoti A, Multini LC, Diouf B, Becker M, Vulcan J, Sylla M, et al. The influence of the larval microbiome on susceptibility to Zika virus is mosquito genotype-dependent. PLoS Pathogens. 2023;**19**(10):e1011727

[66] Acevedo-Guerrero T. *Water with larvae* : Hydrological fertility, inequality, and mosquito urbanism. Environment and Planning E: Nature and Space. 2025;**8**(1):13-30

[67] Goellner E, Neckel A, Bodah BW, Maculan LS, Almeida Silva CCO d, Piccinato D, et al. Geospatial analysis of ae. Aegypti foci in southern Brazil. Journal of Environmental Chemical Engineering. 2021;**9**(6):106645

[68] da Costa PA, Chaves BA, Godoy RSM, Coelho DF, Vieira Júnior AB, Alencar RM, et al. Vector competence for Zika virus changes depending on the Aedes aegypti's region of origin in Manaus: A study of an endemic Brazilian Amazonian City. Viruses. 2023;**15**(3):770

[69] Chouin-Carneiro T, David MR, de Bruycker NF, dos Santos FB, Lourenço-de-Oliveira R. Zika virus transmission by Brazilian Aedes aegypti and Aedes albopictus is virus dose and temperature-dependent. PLoS Neglected Tropical Diseases. 2020;**14**(9):e0008527

[70] Pohlenz TD, Vela J, Reyna M, Fredregill C, Hur B, Erraguntla M, et al. Exposure to Cooler Temperatures During Pupal Development Increases *Aedes aegypti* Vector Competence and the R0 for Zika Virus. 2024

[71] Winokur OC, Main BJ, Nicholson J, Barker CM. Impact of temperature on the extrinsic incubation period of Zika virus in Aedes aegypti. PLoS Neglected Tropical Diseases. 2020;**14**(3):e0008047

[72] Tesla B, Demakovsky LR, Mordecai EA, Ryan SJ, Bonds MH,

Ngonghala CN, et al. Temperature drives Zika virus transmission: Evidence from empirical and mathematical models. Proceedings of the Royal Society B: Biological Sciences. 1884;**2018**(285):20180795

[73] Murrieta RA, Garcia-Luna SM, Murrieta DJ, Halladay G, Young MC, Fauver JR, et al. Impact of extrinsic incubation temperature on natural selection during Zika virus infection of Aedes aegypti and Aedes albopictus. PLoS Pathogens. 2021;**17**(11):e1009433

[74] Carrington LB, Armijos MV, Lambrechts L, Scott TW. Fluctuations at a low mean temperature accelerate dengue virus transmission by Aedes aegypti. PLoS Neglected Tropical Diseases. 2013;**7**(4):e2190

[75] Ruiz-Polo AA, Barrera-Rivera LV. Efecto de la temperatura en el ciclo biológico del Aedes aegypti en condiciones de laboratorio. Revista Peruana de Medicina Experimental y Salud Pública. 2024;**41**(3):327-328

[76] Terradas G, Manzano-Alvarez J, Vanalli C, Werling K, Cattadori IM, Rasgon JL. Temperature Affects Viral Kinetics and Vectorial Capacity of *Aedes aegypti* Mosquitoes Co-Infected with Mayaro and Dengue viruses. 2023

[77] da Costa CF, dos Passos RA, Lima JBP, Roque RA, de Souza SV, Campolina TB, et al. Transovarial transmission of DENV in Aedes aegypti in the Amazon basin: A local model of xenomonitoring. Parasites & Vectors. 2017;**10**(1):249

[78] Ferreira-de-Lima VH, Lima-Camara TN. Natural vertical transmission of dengue virus in Aedes aegypti and Aedes albopictus: A systematic review. Parasites & Vectors. 2018;**11**(1):77

[79] Wang X, Qian C, Zhang C, Hu S, Asad M, Yang C, et al. Zika virus transmission in Aedes aegypti: A systematic study on the ability of mosquitoes to transmit the virus horizontally and vertically. Virologica Sinica. 2025;**40**(2):192-205

[80] Teixeira AF, de Brito BB, Correia TML, Viana AIS, Carvalho JC, da Silva FAF, et al. Simultaneous circulation of Zika, dengue, and chikungunya viruses and their vertical co-transmission among Aedes aegypti. Acta Tropica. 2021;**215**:105819

[81] Lequime S, Paul RE, Lambrechts L. Determinants of arbovirus vertical transmission in mosquitoes. PLoS Pathogens. 2016;**12**(5):e1005548

[82] Danis-Lozano R, Díaz-González EE, Malo-García IR, Rodríguez MH, Ramos-Castañeda J, Juárez-Palma L, et al. Vertical transmission of dengue virus in *Aedes aegypti* and its role in the epidemiological persistence of dengue in central and southern Mexico. Tropical Medicine & International Health. 2019;**24**(11):1311-1319

[83] Thangamani S, Huang J, Hart CE, Guzman H, Tesh RB. Vertical transmission of Zika virus in Aedes aegypti mosquitoes. The American Society of Tropical Medicine and Hygiene. 2016;**95**(5):1169-1173

[84] Joshi V, Mourya DT, Sharma RC. Persistence of dengue-3 virus through transovarial transmission passage in successive generations of Aedes aegypti mosquitoes. The American Journal of Tropical Medicine and Hygiene. 2002;**67**(2):158-161

[85] Buckner EA, Alto BW, Lounibos LP. Larval temperature–food effects on adult mosquito infection and vertical transmission of Dengue-1

virus. Journal of Medical Entomology. 2016;**53**(1):91-98

[86] Martins VEP, Alencar CH, Kamimura MT, de Carvalho Araújo FM, De Simone SG, Dutra RF, et al. Occurrence of natural vertical transmission of Dengue-2 and Dengue-3 viruses in Aedes aegypti and Aedes albopictus in Fortaleza, Ceará, Brazil. PLoS One. 2025;7(7):e41386

[87] Buckner EA, Alto BW, Lounibos LP. Vertical transmission of Key West Dengue-1 virus by Aedes aegypti and Aedes albopictus (Diptera: Culicidae) mosquitoes from Florida. Journal of Medical Entomology. 2013;**50**(6):1291-1297

[88] Edillo FE, Sarcos JR, Sayson SL. Natural vertical transmission of dengue viruses in *Aedes aegypti* in selected sites in Cebu City, Philippines. Journal of Vector Ecology. 2015;**40**(2):282-291

[89] Izquierdo-Suzán M, Zárate S, Torres-Flores J, Correa-Morales F, González-Acosta C, Sevilla-Reyes EE, et al. Natural vertical transmission of Zika virus in larval *Aedes aegypti* populations, Morelos, Mexico. Emerging Infectious Diseases. 2019;**25**(8):1477-1484

[90] Costa CF d, Silva AV d, Nascimento VA d, Souza VC, Monteiro DC d S, Terrazas WCM, et al. Evidence of vertical transmission of Zika virus in field-collected eggs of Aedes aegypti in the Brazilian Amazon. PLoS Neglected Tropical Diseases. 2018;**12**(7):e0006594

[91] Chaves BA, Junior ABV, Silveira KRD, Paz A d C, Vaz EB d C, Araujo RGP, et al. Vertical transmission of Zika virus (Flaviviridae, Flavivirus) in Amazonian Aedes aegypti (Diptera: Culicidae) delays egg hatching and larval development of progeny. Journal of Medical Entomology. 2019;**56**(6):1739-1744

[92] Comeau G, Zinna RA, Scott T, Ernst K, Walker K, Carrière Y, et al. Vertical transmission of Zika virus in Aedes aegypti produces potentially infectious progeny. The American Journal of Tropical Medicine and Hygiene. 2020;**103**(2):876-883

[93] Nag DK, Payne AF, Dieme C, Ciota AT, Kramer LD. Zika virus infects Aedes aegypti ovaries. Virology. 2021;**561**:58-64

[94] Ciota AT, Bialosuknia SM, Ehrbar DJ, Kramer LD. Vertical transmission of Zika virus by *Aedes aegypti* and *ae Albopictus* mosquitoes. Emerging Infectious Diseases. 2017;**23**(5):880-882

[95] Maimusa HA, Ahmad AH, Kassim NFA, Rahim J. Age-stage, two-sex life table characteristics of *Aedes albopictus* and *Aedes Aegypti* in Penang Island, Malaysia. Journal of the American Mosquito Control Association. 2016;**32**(1):1-11

[96] Wohl MP, McMeniman CJ. Overview of *Aedes aegypti* and use in laboratory studies. Cold Spring Harbor Protocols. 2023;**2023**(3):pdb.top107651

[97] Relyea RA. The impact of insecticides and herbicides on the biodiversity and productivity of aquatic communities. Ecological Applications. 2005;**15**(2):618-627

[98] Benelli G. Plant-borne ovicides in the fight against mosquito vectors of medical and veterinary importance: A systematic review. Parasitology Research. 2015;**114**(9):3201-3212

[99] Pavela R. Essential oils for the development of eco-friendly mosquito larvicides: A review. Industrial Crops and Products. 2015;**76**:174-187

[100] Pavela R. History, presence and perspective of using plant extracts as

commercial botanical insecticides and farm products for protection against insects - a review. Plant Protection Science. 2016;**52**(4):229-241

[101] Silvério MRS, Espindola LS, Lopes NP, Vieira PC. Plant natural products for the control of Aedes aegypti: The Main vector of important arboviruses. Molecules. 2020;**25**(15):3484

[102] Priya SS, Vasantha-Srinivasan P, Altemimi AB, Keerthana R, Radhakrishnan N, Senthil-Nathan S, et al. Bioactive molecules derived from plants in managing dengue vector Aedes aegypti (Linn.). Molecules. 2023;**28**(5):2386

[103] Fernandes DA, Rique HL, Oliveira LHG d, Santos WGS, Souza M d FV d, Nunes F d. C. Ovicidal, pupicidal, adulticidal, and repellent activity of Helicteres velutina K. Schum against Aedes aegypti L. (Diptera: Culicidae). Brazilian Journal of Veterinary Medicine. 2021;**43**:e112120

[104] Alievi K, Capoani GT, Buzatto M, Miorando D, Serpa PZ, Fogolari O, et al. Ateleia glazioveana and Ocimum basilicum: Plants with potential larvicidal and repellent against Aedes aegypti (Diptera, Culicidae). Research, Society and Development. 2021;**10**(17):e228101724733

[105] Torawane S, Andhale R, Pandit R, Mokat D, Phuge S. Screening of some weed extracts for ovicidal and larvicidal activities against dengue vector Aedes aegypti. The Journal of Basic and Applied Zoology. 2021;**82**(1):36

[106] Yagoo A, Milton MCJ, Vilvest J, Johnson I, Balakrishna K. Mosquito larvicidal, pupicidal and ovidical effects of the different extracts of the leaves of Peltophorum pterocarpum against Aedes aegypti and Culex quinquefasciatus. Future Journal of Pharmaceutical Sciences. 2023;**9**(1):32

[107] Yagoo A, Milton MCJ, Vilvest J. Mosquito larvicidal, pupicidal and ovicidal effects of the extracts of Peltophorum pterocarpum flowers on Aedes aegypti and Culex quinquefasciatus. Veterinary Parasitology: Regional Studies and Reports. 2023;**46**:100929

[108] Silva JR d A, Oliveira AA d, França LP, da Cruz JD, ACF A. Exploring the Larvicidal and Adulticidal activity against Aedes aegypti of essential oil from Bocageopsis multiflora. Molecules. 2024;**29**(10):2240

[109] Selvakumaran J, Ragavendran K, Ignacimuthu S, Sivanandhan S, Reegan AD, Aremu AO, et al. Mosquitocidal susceptibility and non-target effects of essential oil from Brassica nigra W.D.J.Koch seeds on immature stages of Aedes aegypti L., Anopheles stephensi Liston and Culex quinquefasciatus say. South African Journal of Botany. 2024;**167**:578-584

[110] Maung Mya M, Thura Lin SH, Win Oo A, Htay TM, Nwe CT, Nyein Mu TZ, et al. Larvicidal, ovicidal and repellency properties of the indigenous fruit Citrus reticulata (Orange) peels essential oil against Aedes aegypti mosquitoes. Middle East Journal of Applied Science & Technology. 2024;**07**(04):66-80

[111] Omoniwa BP, Johnson TO, Acho MA, Nwonuma CO, Soji-Omoniwa O, Oluwafemi AG, et al. Larvicidal and adulticidal activities of essential oil of Helianthus annuus (L.) seed and molecular docking of its GCMS constituents against Aedes aegypti acetylcholinesterase. Informatics in Medicine Unlocked. 2024;**45**:101446

[112] Lim V, Mohd Narawi M, Chiu HI, Tung WH, Tan JJ, Lee CK. Selected essential oils as repellents against *Aedes Aegypti* : Validation of the bioconstituents

using gas chromatography. Journal of Essential Oil-Bearing Plants. 2019;**22**(4):1058-1073

[113] Bursali F. Larvicidal and ovicidal effects of selected chemicals found in plant-derived essential oils for mosquito control. Turkish Bulletin of Hygiene and Experimental Biology. 2024;**81**(3):287-296

[114] Gültekin ZN, Kahraman Ş, Çetin H. Larvicidal and Ovicidal properties of essential oils derived from Origanum minutiflorum and salvia dorystoechas against the Aedes aegypti. Commagene Journal of Biology. 2024;**8**(1):18-23

[115] Almadiy AA. Chemical composition, insecticidal and biochemical effects of two plant oils and their major fractions against Aedes aegypti, the common vector of dengue fever. Heliyon. 2020;**6**(9):e04915

[116] Soonwera M, Moungthipmalai T, Aungtikun J, Sittichok S. Combinations of plant essential oils and their major compositions inducing mortality and morphological abnormality of Aedes aegypti and Aedes albopictus. Heliyon. 2022;**8**(5):e09346

[117] Zaniol F, Calisto JFF, Cozzer G, Ferro DM, Dias JL, Rodrigues LGG, et al. Comparative larvicidal effect of Pterodon spp. extracts obtained by different extraction methods. Journal of Supercritical Fluids. 2020;**166**:104993

[118] Spinozzi E, Ferrati M, Giudice DL, Felicioni E, Petrelli R, Benelli G, et al. Microwave-assisted Hydrodistillation of the insecticidal essential oil from Carlina acaulis: A fractional factorial design optimization study. Plants. 2023;**12**(3):622